T0309395

IN SEARCH OF

BIOHAPPINESS

Biodiversity and Food, Health and Livelihood Security

2nd Edition

IN SEARCH OF BIOHAPPINESS

Biodiversity and Food, Health and Livelihood Security

2nd Edition

M S Swaminathan

M S Swaminathan Research Foundation, India

NEW JERSEY • LONDON • SINGAPORE • BEIJING • SHANGHAI • HONG KONG • TAIPEI • CHENNAI

Published by

World Scientific Publishing Co. Pte. Ltd.

5 Toh Tuck Link, Singapore 596224

USA office: 27 Warren Street, Suite 401-402, Hackensack, NJ 07601

UK office: 57 Shelton Street, Covent Garden, London WC2H 9HE

Library of Congress Cataloging-in-Publication Data

Swaminathan, M. S. (Monkombu Sambasivan).

In search of biohappiness : biodiversity and food, health and livelihood security / by M.S. Swaminathan (M.S. Swaminathan Research Foundation, India). -- 2nd edition.

pages cm

Includes index.

ISBN 978-9814656931

1. Biodiversity conservation. 2. Biodiversity. 3. Food security. 4. Food supply. 5. Sustainable agriculture. I. Title.

QH75.S813 2015

333.95'16--dc23

2014049362

British Library Cataloguing-in-Publication Data

A catalogue record for this book is available from the British Library.

In-house Editors: Dipasri Sardar/Amanda Yun

Typeset by Stallion Press

Email: enquiries@stallionpress.com

Printed in Singapore

Contents

Preface

India is a biodiversity-rich nation and biodiversity is the feedstock for the breeding and biotechnology enterprises. As explained in the first edition of this book, biohappiness results from the conservation and sustainable and equitable use of biodiversity. The pathway to biohappiness lies in converting biodiversity hot spots into happy spots. Also, it requires a considerable degree of additional attention in bridging the gap between scientific 'know-how' and field level 'do-how'. It also requires greater investment in public good research by organisations committed to harnessing science for public good. I would like to briefly indicate what needs to be done both in the area of bridging the know-how:do-how gap and in supporting public good research.

Scientific Know-How and Field Level Do-How

Recently, addressing the scientists of the Indian Council of Agricultural Research (ICAR), Prime Minister Narendra Modi emphasised the need to initiate on a large scale a 'lab to land' programme on the lines I had organised in the 1970s. The reason for this exhortation is the growing gap between laboratory discoveries and their field application, resulting in the low productivity of major crops. One area where we observe a wide gap between knowledge and its field application is in the area of biotechnology. For example, many farmers do not grow the old traditional varieties along with new Bt cotton hybrids, although scientists recommend the creation of a refuge to avoid the origin of more virulent races of pests. To effectively accomplish the goal of bridging the know-how:do-how gap, we need to pay attention to the following four kinds of linkages.

(1) Lab to lab: This involves collaboration among all the institutions working on a particular problem like the development of new varieties of wheat and rice, among others. This kind of beneficial partnership is being achieved through the all-India coordinated research projects of the ICAR. Such coordinated projects help to bring together scientists working in agriculture universities, ICAR institutes, and often the private sector too, to undertake the evaluation of new varieties not only from the point of view of yield, but also from the point of view of resistance to pests and diseases as well as to drought, floods and other abiotic stresses. The rapid progress which could be made in the early years of the Green Revolution in the 1960s was only because we could mobilise the power of partnership among scientists working under different agro-climatic and socio-economic conditions. This method of organising all the relevant institutions for undertaking joint varietal evaluation and release is one of the important factors responsible for our rapid progress in improving food production.

(2) Lab to land: The lab to land programme took the form of national demonstrations as well as the establishment of Farm Science Centres, known as *Krishi Vigyan Kendras* (KVKs). There are now several hundred KVKs spread all over the country. These are attached to agriculture universities and research institutions which generate new technologies. The first KVK was established at Puducherry in the 1970s. These institutions, which impart location-specific knowledge, have become powerful instruments for the knowledge and skill empowerment of small farm families. Special attention is given to women farmers as well as to women in farming. As a result of such farmer-scientist collaboration, farmers converted a small government programme into a mass movement, leading to the Green Revolution.

(3) Land to lab: It is clear that there is considerable amount of traditional knowledge and wisdom available with farm families. Such knowledge and ecological prudence arise from practical experience. It is said that "one ounce of practice is worth tonnes of theory". Therefore, we should pay attention to farmers' own traditional practices and varieties. Fortunately, the Protection of Plant Varieties and Farmers' Rights Authority (PPVFRA) has instituted Genome Saviour awards for recognising and rewarding the invaluable contributions of rural and tribal people to genetic resources conservation and enhancement. A good database is also being maintained on traditional knowledge and technologies at the Indian Institute of Management, Ahmedabad, under the guidance of professor Anil Gupta.

PPVFRA recognises the multiple roles of farmers. They are cultivators and guardians of the food security of the country not only because they are the ones who produce the food, but because they also play an important intellectual role in agro-biodiversity, plant selection, plant hybridisation, and so on. In fact, in China, it was the farmers of Hainan Island who identified the male sterile plants which provided the basis for hybrid rice. Sadly, however, farmers' intellectual role in the evolution of new varieties and new technologies is usually unrecognised. This is why in the FAO forum, we asked for farmers' rights, not only breeders' rights. The Union for the Protection of New Varieties of Crops (UPOV), located in Geneva, along with the World Intellectual Property Rights Organisation (WIPO), recognises the intellectual property rights of plant breeders. I have been suggesting that they should recognise the rights of both breeders and farmers, but this has not yet happened. I am confident that someday UPOV will become the International Union for the Protection of Breeders' and Farmers' Rights. In India, we decided to emphasise farmers' rights alongside breeders' rights in the formulation of the Protection of Plant Varieties and Farmers' Rights Act, 2001. This is the only legislation in the world which concurrently recognises farmers' and breeders' rights.

(4) Land to land: This involves farmer-to-farmer learning. The National Commission of Farmers (NCF) had laid particular stress on promoting farmer-to-farmer learning through Farm Schools. Fortunately, industrial houses and financial institutions are promoting the establishment of Farm Schools under their corporate social responsibility programmes. For example, the M.S. Swaminathan Research Foundation has established seven Farm Schools with financial support from the Indian Overseas Bank. Farmer-to-farmer learning is an effective method of technology transfer due to the sound economic basis which guides farmers' decision making.

Land is a shrinking resource for agriculture and we have no option except to produce more from diminishing per capita land and water resources. It is in this context that bridging the know-how:do-how gap becomes both urgent and essential. Our Prime Minister has done a great service by emphasising the need for agricultural research institutions and farm universities to intensify lab to land and other associated scientist-farmer technology sharing programmes.

Public Good Research in Agriculture

Public good research in agriculture is designed to promote risk-minimising agronomy as well as ease of adoption by small and marginal farmers. For example, public good research institutions concentrate on the development of varieties rather than hybrids, since in the case of hybrids, the farmer has to buy the seed every year from the company. In contrast, farmers can keep their own seeds of good varieties of crops. We should not underestimate the power of public good research in contrast to profit-maximising private sector research. I shall try to illustrate this with two recent examples, one dealing with Basmati rice and another, with semi-dwarf wheat varieties.

Basmati is appropriately referred to as the Queen of Rice and has been cultivated for centuries in the foothills of the Himalayas. Because of its culinary quality, it is valued highly in national and international markets. Pakistan is a major producer of Basmati rice. After the advent of the high-yielding varieties of rice that possess genes for semi-dwarf character, Basmati rice was given less importance due to its tall stature and low yield potential (about one tonne per hectare). It is to overcome this difficulty that in 1965 I started breeding semi-dwarf Basmati strains which could respond well to fertiliser and irrigation water application, at the Indian Agricultural Research Institute (IARI). This work which began over 40 years ago has now resulted in outstanding varieties like Pusa Basmati 1121 which has helped to increase foreign exchange earnings by Rs 33,000 crore during 2013–14. During 2014, IARI released another variety Pusa Basmati 1509 which matures in 120 days and has dwarf stature, sturdy stem, non-lodging and non-shattering habits. Also, Basmati 1509 yields about 5 tonnes per hectare. Because of its early maturity, Pusa Basmati 1509 has become very popular among farmers adopting a rice-wheat rotation. No wonder, there is enormous demand for the seeds of this variety which is now occupying a major portion of the Basmati area. This is a good example of the power of public good research, with sharp focus on increasing the productivity and profitability of small holdings.

As in the case of Pusa Basmati rice, IARI has also been developing and releasing wheat varieties which help to increase the production and productivity of this crop. Starting with the semi-dwarf wheat varieties like Kalyan Sona and Sonalika in the 1960s (these are selections made from the material sent by Dr. Norman Borlaug), the Pusa wheats as well as those bred by the scientists of the Punjab Agricultural University and other agricultural universities and institutions have transformed our wheat scenario. From about 7 million tonnes in 1947, wheat production has now reached about 96 million tonnes. The aim is to produce within the next twenty years 150 million tonnes of wheat in 30 million hectares. This will be possible, considering the fact that our wheat breeders are continually producing outstanding new strains. A recent example is HD 2967 developed by the wheat breeders of IARI. This variety now occupies about 6 million hectares in north India. It is resistant to major pests and diseases and yields on an average over 4 tonnes per hectare. It is only this kind of research which can help us achieve an evergreen revolution leading to the enhancement in productivity in perpetuity without ecological harm. The approximate value of the total production of HD 2967 is Rs 45,500 crore. The variety has contributed about 35 million tonnes of wheat.

While the benefits from wheat and rice research have been phenomenal both to the farmers and to the country, the actual expenditure on such research has been only a few crore rupees per year. Thus, the return from investment in public good research is exceedingly impressive. This is why the National Commission on Farmers has laid emphasis on adequate support for research and training at our national research institutions and agriculture universities. This is the best investment the country can make in the interest of sustainable food security, thereby making possible the effective implementation of the provisions of the National Food Security Act.

Among the facilities needed for strengthening public good research are translational research centres which can convert laboratory findings into field application. We also need facilities for taking more than one crop per year, such as greenhouses and growth chambers. In the case of wheat, accelerated work became possible because of the establishment of a centre at Wellington in the Nilgiri Hills where a summer crop of wheat can be taken. This is also possible in the Lahul and Spiti Valley region of Himachal Pradesh. In the case of rice, IARI and Tamil Nadu Agricultural University (TNAU) have established an off-season multiplication centre at Aduthurai in the Thanjavur district of Tamil Nadu. This again helps to purchase time in breeding. We need to augment such facilities so that we will be able to meet the new challenges arising from climate change and global warming. There is no time to relax on the food production front. Eternal vigilance is the price of stable agriculture.

2015 marks the beginning of the UN Decade for Sustainable Development. Sustainable food and nutrition security as well as livelihood security will depend on our efforts to conserve biodiversity. I hope this book will stimulate interest in launching an era of biohappiness. As recognised by the Kingdom of Bhutan on the visionary suggestion of His Majesty the King, the concept of Gross National Happiness (GNH) should replace Gross Domestic Product (GDP) as the measurement of progress. Biohappiness is essential for promoting GNH. I hope this book will make a contribution to achieving the goal of GNH.

I am indebted to Ms. Gita Gopalkrishnan and Ms. Dilhara Begam for their invaluable help in compiling the papers constituting this book. My sincere thanks also go to Ms. Ranjana Rajan and Ms. Dipasri Sardar of Academic Consulting and Editorial Services (World Scientific Group) for their assistance in attending to the details of the publication.

M.S. Swaminathan

Foreword

One fundamental fact of life is not always apparent: People are species, and our life depends on maintaining healthy ecosystems that provide nutritious food to eat, pure water to drink, and clean air to breathe. "Biohappiness", by one of the world's leaders in food production systems, makes this point very clearly. The book provides both practical and ethical reasons why a productive future depends on conserving biodiversity, using biological resources sustainably, and equitably sharing the benefits arising from the use of genetic resources.

Governments, conservation organisations such as my own, and many businesses are now working together toward such a future, based on the Convention on Biological Diversity (CBD). "Biohappiness" gives particular attention to the importance of agricultural lands in achieving the objectives of the CBD, especially through the knowledge that has grown through many generations of farmers working in the world's rural settings. They have developed cropping systems appropriate to their local conditions and have recognised the value of a diversity of crops and varieties of them. This knowledge is often held by indigenous peoples, and such traditional knowledge is shown in this book to have particular value at a time when food production is becoming a growing challenge.

The book illustrates through numerous examples, especially from India, that rural people often grow crop varieties that contain genes that are valuable to crop breeders who are seeking to enhance yields or build greater pest resistance into the major food crops. It cites the CBD for recognizing this expression of traditional knowledge, and indicates that many organisations are now adopting ways to ensure that the holders of such knowledge are appropriately compensated for its wider application. The CBD's Nagoya Protocol on Access and Benefit Sharing, which entered into force in October 2014, provides the framework for ensuring that the world has access to genetic resources and that those providing these resources receive appropriate compensation. Many of the ideas for this Protocol were strongly influenced by the work of Dr. M.S. Swaminathan, and are well reflected in this book.

A productive future for humanity may well depend on this becoming *The Biological Century*, a time when new technologies (such as functional genomics), stronger efforts to conserve biodiversity (including through establishing well-managed protected areas that deliver multiple benefits to the surrounding lands), more inclusive economics (for example, recognising the full value of ecosystem services), and a better appreciation of the diversity of knowledge systems combine to lead to a better life for all: Biohappiness.

My own organisation, the International Union for Conservation of Nature, strongly supports Dr. M.S. Swaminathan's vision as presented in this remarkable book. We are proud of the fact that he was President of IUCN from 1984 to 1990. We continue to work closely with the CBD and other such instruments that are seeking to maintain healthy ecosystems, the diversity of life, greater equity, and the diversity of knowledge systems. We warmly welcome the kinds of biopartnerships that this book seeks to generate, and look forward to joining forces with the growing number of supporters of Biohappiness.

Julia Marton-Lefèvre
Director General
International Union for Conservation of Nature (IUCN)
Gland, Switzerland
21 December 2014

Introduction

How can we define biohappiness? I would say that it is the sustainable and equitable use of biodiversity leading to the creation of more jobs and income. When the use of biodiversity leads to sustainable livelihood security, the local population develops an economic stake in conservation. It means that growth and progress must be reliable and dependable and maintained at an even and steady pace. In farming, it is the production of high yields *in perpetuity*, without associated social or ecological harm. Sustainable development must be firmly rooted in the principles of ecology, social and gender equity, employment generation, and economic advance.

Biodiversity provides the building blocks for sustainable food, health and livelihood security systems. It is the feedstock for both the biotechnology industry and a climate-resilient farming system. Because of its importance for human well-being and survival, a Convention on Biological Diversity (CBD) was adopted at the UN Conference on Environment and Development held at Rio de Janeiro in 1992. The Convention's triple goals are: conservation, sustainable use, and equitable sharing of benefits. CBD defines biological diversity as follows:

> Variability among living organisms from all sources including, *inter alia,* terrestrial, marine and other aquatic ecosystems and the ecological complexes of which they are part; this includes diversity within species, between species and of ecosystems.

The Convention stresses the need for respecting, preserving and maintaining the knowledge, innovations, and practices of indigenous and local communities embodying traditional lifestyles relevant for the conservation and sustainable use of biological diversity. It also calls for the promotion of their wider application with the approval and involvement of the holders of such knowledge, innovations, and practices and for the equitable sharing of the benefits arising from the utilisation of such knowledge, innovations, and practices Article 8(j).

The Convention also recognises that the biodiversity existing within a country is the sovereign property of its people. India is a signatory to CBD and has enacted a National Biodiversity Act which has been in force since 2002. India is classified as a mega biodiversity area from the point of view of species richness and agro-biodiversity. However, two of the major biodiversity rich areas, northeast India and the Western Ghats region, are also classified as "hot spot" areas from the point of view of threats to biodiversity.

In spite of the importance given to the conservation of biodiversity, genetic erosion is progressing in an unabated manner, both globally and nationally. For example, 12 per cent of birds, 21 per cent of mammals, 30 per cent of amphibians, 27 per cent of coral reefs and 35 per cent of conifers and cycads are currently facing extinction. According to the World Conservation Union (IUCN), over 47,677 species may soon disappear. A comprehensive study published in *Science* (29 April 2010) has revealed that there has been no notable decrease in the rate of biodiversity loss between 1970 and 2010. Even a very unique species like the orang-utan, the closest relative of man, is threatened with extinction in the Island of Borneo.

To generate awareness of the urgency of genetic resources conservation, 22 May of every year is being commemorated as the International Day for Biological Diversity. 2010 has also been designated by the United Nations as the International Year of Biodiversity. The challenge now is for every country to develop an implementable strategy for saving rare, endangered and threatened species through education, social mobilisation and regulation. Meaningful results can be obtained only if biodiversity conservation is considered in the context of sustainable development and poverty alleviation. Indira Gandhi pointed out at the UN Conference on the Human Environment held at Stockholm in 1972 that unless we attend concurrently to the needs of the poor and of the environment, the task of saving our environmental assets will not be easy. Biodiversity loss is predominantly related to habitat destruction largely for commercial exploitation as well as for alternative uses like roads, buildings, etc. Invasive alien species and unsustainable development are other important causes of genetic erosion. How can we reverse the paradigm and enlist development as an effective instrument for conserving biodiversity? Let me cite a few examples to illustrate how biodiversity conservation and development can become mutually reinforcing.

In 1990, I visited MGR Nagar, a village near Pichavaram in Tamil Nadu, for studying the Mangrove Forests of that area. The families living in MGR Nagar were extremely poor and were still waiting for the benefits of government schemes to be allocated to them. The children had no opportunities for education, the schools were far away and getting admission was difficult. I then told my colleagues that saving mangrove forests without saving the children for whose well-being these forests were being saved made no sense. With the help of a few donors, we started a primary school in the village for all the children, irrespective of their age. A few years later, the State government took over the school and expanded its facilities. It now needs to be upgraded into a higher secondary school. After the 2004 tsunami, the hutment dwellings have been replaced by brick houses and the whole scenario of MGR Nagar has totally changed. During the tsunami, the mangroves served as speed breakers and saved the people of the village from the fury of the tidal waves. Everyone in the village now understands the symbiotic relationship between mangroves and coastal communities, that the root exudate from the mangrove trees enriches the water with nutrients and promotes sustainable fisheries. It is clear that hereafter mangroves in this region will be in safe hands.

Another example relates to the tribal families of Kolli Hills in Tamil Nadu. The local tribal population had been cultivating and conserving a wide range of millets and medicinal plants. However, due to lack of market demand for traditional foods, they had to shift to more remunerative crops like tapioca and pineapple. The millet crops cultivated and consumed by them for centuries were rich in protein and micronutrients. They were also much more climate resilient, since mixed cropping of millets and legumes minimises risks arising from unfavourable rainfall. Such risk distribution agronomy is the saviour of food security in an era of climate change. How then can we revitalise the conservation traditions of tribal families, without compromising on their economic well-being? M.S. Swaminathan Research Foundation (MSSRF) scientists started a programme designed to create an economic stake in conservation, by both value addition to primary products and by finding niche markets for their traditional foodgrains. Commercialisation thus became the trigger for conservation. Today many of the traditional millets are once again being grown and consumed. They now proudly sing, "Biodiversity is our life", which is also the key message of the International Year for Biodiversity.

A third example relates to the tribal areas of the Koraput region of Odisha (formerly Orissa), which is an important centre of diversity of rice. Fifty years ago, there were over 3500 varieties of rice in this area. Today this number has come down to about 300. Even with these three hundred varieties, it is essential that the tribal families derive some economic benefit from the preservation of such rich genetic variability in rice. Now, they, in partnership with scientists, have developed improved varieties like *Kalinga Kalajeera*, which fetches a premium price in the market. For too long, tribal and rural families have been conserving genetic resources for public good at personal cost. It is time that we recognise the importance of promoting a genetic conservation continuum, starting with the simple situation of *in situ* on-farm conservation of landraces by local communities, and extending to technological breakthroughs like the preservation of a sample of genetic variability under permafrost conditions at locations like Svalbard near the North Pole maintained by the Government of Norway or Chang La in Ladakh where our Defence Research and Development Organisation has established a conservation facility.

While giving operational content to the concept of sustainable development, we should ask some of the questions Dudley Sears asked decades ago[1]:

> The questions to ask about a country's development are: What has been happening to poverty? What has been happening to unemployment? What has been happening to inequality? If all three of these have become less severe, then beyond doubt this has been a period of development for the country concerned. If one or two of these central problems have been growing worse, especially if all three have, it would be strange to call the result "development", even if per capita income doubled.

In dealing with issues relating to biodiversity, we should also ask similar questions. Is the biodiversity management system conducive to the reduction of poverty, promotion of gender equity, and the generation of livelihood opportunities? The need for gender and social equity in sharing benefits from the commercial use of biodiversity cannot be overemphasised, if we are to succeed in preventing further genetic erosion. Biodiversity rich countries are characterised by agro-ecological variability. In addition, there is a strong positive correlation between cultural diversity and agro-biodiversity. Women, in particular, tend to conserve and improve plants of value in strengthening household nutrition and health security. It is, therefore, imperative to give explicit recognition to the role of women in genetic resource conservation and enhancement.

How can we harness biodiversity for poverty alleviation? Obviously, this can be done only if we can convert biodiversity into jobs and income on a sustainable basis. Several institutional mechanisms have been developed at MSSRF for this purpose, such as biovillages and biovalleys. In biovillages, the conservation and enhancement of natural resources like land, water and biodiversity become priority tasks. At the same time, the biovillage community aims to increase the productivity and profitability of small farms and

[1]Sears, D., 1969. "The Meaning of Development". *International Development Review* 11(4).

create new livelihood opportunities in the non-farm sector. Habitat conservation is vital for preventing genetic erosion. In a biovalley, the local communities try to link biodiversity, biotechnology and business in a mutually-reinforcing manner. For example, the herbal biovalley under development in Koraput aims to conserve medicinal plants and local foods and convert them into value-added products based on assured and remunerative market linkages. Tribal families in Koraput have formed a "Biohappiness Society".

There is need for launching a Biodiversity Literacy Movement, so that right from childhood everyone is aware of the importance of diversity for the maintenance of food, water, health and livelihood security as well as a climate-resilient food production system. The Government of India has started programmes like Deoxyribonucleic Acid (DNA) and Genome Clubs to sensitise schoolchildren about the importance of conserving biodiversity. The Government has also started recognising and rewarding the contributions of rural and tribal families in the field of genetic resources conservation through Genome Saviour Awards. We need similar awards for those who are conserving breeds of animals, forests, and fishes.

The Biodiversity Act envisages action at three levels: the Panchayat Biodiversity Committee (responsible for conservation as well as for operationalising the concept of prior informed consent and benefit sharing), the State Biodiversity Board and the National Biodiversity Authority. These three units of the bioresources conservation movement should ensure that all development programmes are subjected to a **biodiversity impact** analysis, so that economic advance is not linked to biodiversity loss. The Biodiversity Day and the Biodiversity Year remind us that we should do everything possible to spread bioliteracy among the public and usher in an era of biohappiness in biodiversity rich areas. Then, "biodiversity hot spots" will become "biodiversity happy spots". This should be our commitment to the generations yet to be born.

Section I

Conservation, Cultivation, Consumption and Commerce: Pathways to Biohappiness

Chapter 1

Towards an Era of Biohappiness

The global food security situation is entering a critical phase. International prices of wheat, rice, maize and other food crops are going up due to the gap between demand and supply. Petroleum prices are going up steeply. As a consequence, there is diversion of both farm land and grains for fuel production. The State of Iowa in USA, which used to be known as a State that feeds the world, is now proud of calling itself as the State that fuels the world. Compounding these factors is the growing threat of climate change resulting in more frequent drought, floods and pest epidemics. Further there is a danger to coastal agriculture and communities arising from sea level rise. It is in this context that the conservation and sustainable and equitable use of biodiversity assume urgency. It would therefore be useful to consider some of the major issues relating to biodiversity conservation and use.

Biodiversity conservation is a continuum. Two ends of this continuum, namely *in situ* conservation and *ex situ* preservation, receive both political and public attention and support. However, the vast amount of *in situ* on-farm conservation work being carried out by tribal and rural women and men remains largely unrecognised and unrewarded. It is this segment of conservation of genetic diversity which contributes significantly to food and health security. This is the most value-added component of biodiversity conservation, since the material conserved by local communities undergoes both selection for desirable qualities and knowledge addition through observation, experimentation and experience. Yet, this component received little attention or appreciation until the Food and Agricultural Organisation (FAO) promoted the concept of Farmers' Rights and the Convention on Biological Diversity (CBD) accorded explicit recognition to the conservation traditions of tribal and rural families.

Article 8(j) of the CBD calls on the contracting parties to respect, preserve and maintain the knowledge, innovations and practices of indigenous and local communities embodying traditional lifestyles. It also calls for the equitable sharing of benefits arising from the utilisation of such knowledge, innovations and practices. The absence of an internationally-agreed methodology for sharing economic benefits from the commercial exploitation of biodiversity with the primary conservers and holders of traditional knowledge and information is leading to a growing number of accusations of biopiracy committed by business and industry in developing countries. Biopiracy can be converted into bio-partnership only if the principles enshrined in Article 8(j) of CBD are adopted both in letter and in spirit by public and private sector institutions and commercial enterprises.

Equity in benefit sharing is fundamental to the retention and revitalisation of the *in situ* on-farm conservation traditions of rural and tribal families. Material and information transfer agreements should safeguard the interests of those providing the concerned material/information. The institutions that fall under The Consultative Group on International Agricultural Research (CGIAR) are already adopting a Material Transfer Agreement which will prevent the monopolistic exploitation of public-funded research on Plant Genetic Resources (PGR) for commercial profit. Benefit sharing procedures will have to be developed at the individual and community levels. The same procedures for seeking recognition and reward as those available to professional breeders can be used at the level of an individual farmer conserver/innovator, with assistance provided in obtaining patents/plant variety protection in accordance with the prescribed national legislation. The problem is more complex in the case of benefit sharing with entire communities. Procedures are available for identifying the area from which critical genes responsible for the commercial success of a new variety have come. Thanks to molecular techniques, this possibility also extends to genes controlling quantitative traits like yield and quality. Therefore, appropriate reward can be given from the Community Biodiversity and Gene Funds proposed to be established under Biodiversity and Plant Variety Protection Acts in several developing countries. Breeders will have to be requested to disclose the full pedigrees of their new varieties and indicate to the extent possible the area from where the critical genes, including Quantitative Trait Loci (QTL) have been accessed. The communities concerned can decide on the use of the funds provided. Obviously they should be used for community benefits, including the funds needed for strengthening on-farm conservation of landraces and seed storage and technology.

The Indian Legislation

India is so far the only country which has enacted a law to accord concurrent recognition to the rights of breeders and farmers. The Indian Plant Variety Protection and Farmers' Rights Act contains provisions for according recognition and reward to farmers/farm communities from the National Gene Fund in recognition of their invaluable contributions to the conservation of genetic resources and their improvement through selection, preservation and knowledge addition.

The preamble to the Act calls for recognition to the contributions of farm families to crop improvement made at any time. Agrobiodiversity centres like the Koraput Region in Odisha where tribal families have preserved and improved rice genetic material over many centuries need to be protected from genetic erosion. The tribal families who have conserved important genetic material for public good at personal cost were recently honoured by the Government of India with the first Genome Saviour Award.

The important criterion to be used in the exercise relating to recognition and reward will be the role played by the landrace/ farmers' variety in the breeding of improved varieties either in the public or private sector, with specific characteristics derived from the farmers' variety. The role of the landrace in providing the needed critical gene(s) in the development of the new variety will have to be indicated in the pedigree to be provided by the breeder at the time of the registration of the variety. For example, *Oryza nivara* from eastern Uttar Pradesh provided the genes for tungro resistance in the very widely grown rice variety IR36. Similarly, a farmers' cotton variety, *Bikaneri Nerma*, has proved to be the most important parent in several Bt cotton hybrids.

It may not be always possible to attribute the donor landrace to an individual farmer. In such instances, the recognition will go to the community, which has preserved such a valuable germplasm. The recognition can include conferring titles to individuals/communities for undertaking conservation over a long period. The reward may include a substantial monetary amount determined on the basis of the contribution of critical genes and continuity of conservation at personal/community cost. The cost of recognition and reward may be borne from a specifically earmarked budgetary provision of the National Biodiversity Authority for such purposes or from the National Gene Fund, as the case may be. This reward to the community/farm family may be used for purposes such as strengthening the infrastructure for *in situ* on-farm conservation of landraces and for meeting related needs like community threshing yards and seed storage facilities, etc. When a community is identified for reward, it is important to define its inclusiveness with special care to gendered inclusion and to make sure that the reward flows equitably to the benefit of all and to the cause of conservation.

For example, Super Wheats capable of yielding about 8 t/ha are now in the breeders' assembly line. Such wheats have a complex pedigree and illustrate the importance of genetic resources conservation and exchange. Super Wheats are semi-dwarf with robust stems, broad leaves, large spikes with more number of grains per panicle and more grain weight. The Super Wheat architecture in the breeders' assembly line is derived from a blend of *Tetrastichon* (Yugoslavia), *Agrotriticum* (Canada), *Tetraploid Polonicum* (Poland), *Gigas* (Israel), Morocco wheat (Morocco) and semi-dwarf wheats currently grown in India.

This emphasises the need for the multilateral system of access and benefit sharing enshrined in Articles 10–13 of the International Treaty on Plant Genetic Resources for Food and Agriculture. If Article 9, relating to Farmers' Rights, and Articles 10–13 relating to the multilateral system of the International Treaty are implemented both in letter and spirit the future of biodiversity conservation and sustainable and equitable use will be ensured in perpetuity.

The loss of every species and gene will limit our options for the future, particularly in the context of climate change. Recombinant Deoxyribonucleic Acid (DNA) technology, functional genomics and proteomics and the emerging nanobiotechnology have opened up uncommon opportunities for creating novel genetic combinations of great value to strengthening food, nutrition, health and livelihood security. This is why all the three forms of conservation, namely *in situ, ex situ* and *in situ* on-farm, should be strengthened. The MS Swaminathan Research Foundation (MSSRF) is setting up a biovalley in the Koraput Region of Odisha, which is rich in herbal biodiversity. The biovalley is to biotechnology what the Silicon Valley is to information technology.

Today, commercialisation is leading to overexploitation of habitats rich in biodiversity like rainforests and coral reefs. It is important that we reverse the paradigm and create an economic stake in conservation. It is in this context that the rights of the primary conservers for recognition and reward assume importance. The Indian Plant Varieties and Farmers' Rights Act 2001 is unique in the world in that it combines in one piece the legislation provisions for recognising the rights of both breeders and farmers. Breeders and farmers are allies in the struggle for food and health security and their rights should be mutually reinforcing and not antagonistic. The Indian Act recognises the triple role of a farmer, namely as a cultivator, conserver and breeder. The Genome Saviour Award of the Indian Plant Variety Protection and Farmers' Rights Authority gives explicit recognition to the role of tribal and rural women and men in the field of genetic resources, conservation and enhancement through selection and value-addition through knowledge.

Conservation, cultivation, consumption and commerce should be dealt with in an integrated manner. Public policies should promote the diversification of food habits resulting in the revitalisation of former food traditions which involved the use of a wide range of food plants. Community-level Gene, Seed, Grain and Water Banks should be promoted in order to ensure local-level food and water security. Underutilised or orphan crops can help to eradicate both chronic and hidden hunger. Therefore, the future of our food and health security systems will depend upon our success in making biodiversity conservation everybody's business.

This is the pathway to an era of biohappiness rooted in the principles of ethics and equity in benefit sharing. Biohappiness will help to end the prevailing irony, where the primary conservers remain poor, while those using their knowledge and material become rich. Bio-partnerships leading to biohappiness should guide public policies relating to biodiversity and biotechnology.

Chapter 2

Biodiversity and Sustainable Food Security

Demographic explosion, environment pollution, habitat destruction, enlarging ecological footprint, widespread hunger and unsustainable lifestyles, and potential adverse changes in climate all threaten the future of human food, water, health and livelihood security systems. 2010 appears to mark the beginning of uncertain weather patterns and extreme climate behaviour. Events like temperature rise, drought, flood, coastal storms and rise in sea level are likely to present new challenges to the public, professionals and policy makers. Biodiversity has so far served as the feedstock for sustainable food and health security and can play a similar role in the development of climate-resilient farming and livelihood systems. Biodiversity is also the feedstock for the biotechnology industry. Unfortunately, genetic erosion and species extinction are now occurring at an accelerated pace due to habitat destruction, alien species invasion and spread of agricultural systems characterised by genetic homogeneity. Genetic homogeneity increases genetic vulnerability to biotic and abiotic stresses. To generate widespread interest in biodiversity conservation, the UN General Assembly has declared 2010 as the International Year of Biodiversity.

The Global Convention on Biodiversity (CBD) adopted at the UN Conference on Environment and Development held at Rio de Janeiro in 1992, and the International Treaty on Plant Genetic Resources for Food and Agriculture adopted by Member Nations of the Food and

Agricultural Organisation (FAO) in 2001 provide a road map for the conservation and sustainable and equitable use of biodiversity. CBD emphasises that biodiversity occurring within a nation is the sovereign property of its people. Hence, the primary responsibility for conserving biodiversity, using it sustainably and equitably, and preserving it for posterity rests with each individual nation. This implies that all nations should subject development programmes to a Biodiversity Impact Analysis in order to ensure that economic advance is not linked to biodiversity loss. Inter-generational equity demands that we must preserve for posterity at least a representative sample of the biodiversity existing in our planet today.

Initiatives like the recognition of Globally Important Agricultural Heritage Sites of FAO and the World Heritage Sites of UNESCO are important to generate interest in the conservation and enrichment of unique biodiversity sites. Particular attention will have to be given to sustaining the protected areas through public education and social mobilisation, in addition to appropriate regulation. Unfortunately, many of the protected areas, national parks and Biosphere Reserves are facing serious anthropogenic pressures. Based on the model of the Biosphere Trust for the conservation of the Gulf of Mannar Biosphere Reserve in India developed by MSSRF, Biosphere Reserves could be jointly managed by local communities and government departments. The concept of Participatory Forest Management (PFM) should be extended to national parks and Biosphere Reserves. This will help to foster among the local population the feeling that they are trustees of these unique gifts of nature.

Special attention should be paid to biodiversity hotspots. Through public cooperation, they should be converted into biodiversity "happy spots", where the sustainable use of biodiversity helps to generate new jobs and income. Coastal biodiversity has not received adequate attention. Mangrove wetlands are under various degrees of degradation. The Joint Mangrove Forest Management procedure developed by MSSRF should be implemented wherever mangrove genetic resources still occur. Infrastructure for strengthening community conservation like drying yards, seed storage and seed testing facilities needs to be supported in all agro-biodiversity hotspots.

Biodiversity conservation and sustainable management should become a national ethic. Government agencies including local self-government authorities like panchayats could play an important role in both spreading biodiversity literacy through Community Biodiversity Registers and by creating the necessary infrastructure like Gene and Seed Banks. Awareness of the relationship between biodiversity and human health and farm animal survival should become widespread. Special training programmes should be organised to enable panchayat committees to become well versed with the provisions of the Biodiversity and Protection of Plant Varieties and Farmers' Rights Acts, particularly with those relating to prior informed consent, access and benefit sharing as well as the gene and biodiversity funds.

Women play a lead role in biodiversity conservation and sustainable use. Mainstreaming the gender dimension in all conservation and food security programmes is a must. Women conservers should be enabled to continue their conservation ethos, by providing support for essential infrastructure like seed storages. Agro-biodiversity is the result of interaction between cultural diversity and biodiversity. An important aspect of cultural diversity is culinary diversity. Every step should be taken to recognise and preserve cultural diversity and to blend traditional wisdom with modern science.

The role of farmers and farming in the mitigation of climate change has not so far been adequately recognised and appreciated. Farmers can help build soil carbon banks and at the same time improve soil fertility through fertiliser trees. Mangrove forests are very efficient in carbon sequestration. Biogas plants can help to convert methane emissions into energy for the household. Hence, a movement should be started at global, national and local levels for enabling all farmers with smallholdings and a few farm animals to develop a water-harvesting pond, plant a few fertiliser trees and establish a biogas plant in their farms. I reiterate: just these — a farm pond, some fertiliser trees and a biogas plant — will make every small farm contribute to climate change mitigation, soil health enhancement and water for a crop life-saving irrigation.

To strengthen the linkages between biodiversity and food security, there is a need to enlarge the food basket by including a wide range of millets, tubers and legumes in the diet. Nutrition security can be strengthened by introducing horticultural remedies for nutritional maladies, like the deficiency of iron, iodine, zinc, vitamin A, vitamin B12 and other micronutrients in the diet. Biodiversity in medicinal plants helps to strengthen health security. The role of biodiversity in sustaining livelihoods can be enhanced through crop-livestock-fish integrated farming systems. Bioresources should be converted into jobs and income meaningful to the poor in an environmentally sustainable manner.

Kerala, A Case Study

I would like to describe here the State of Kerala in southwest India which is an agro-biodiversity paradise. Kerala is rich in crops like rice, banana, jackfruit, tubers, spices, medicinal plants, coconut, plantation crops, coastal halophytes, inland and marine fishes, large and small ruminants including the Vechur cow and Malabari and Attapadi goats. The medicinal plant wealth has helped Kerala to perfect the science of ayurveda and thereby become a preferred State for health tourism. The challenge now lies in both preserving and enriching this biological wealth and in converting bio-resources into jobs and income on a sustainable basis.

Kerala has a long tradition of *in situ* on-farm conservation in crops like rice, spices and tubers as well as *ex situ* preservation through sacred groves, botanical gardens, Biosphere Reserves and aquaria. Tribal communities have conserved life-saving crops particularly tubers and medicinal plants, and traditional healers have deep knowledge of the therapeutic value of local flora. Speciality rices like *Njavara* have been identified and conserved. Farmers have been serving as conservers, breeders and cultivators. Several important varieties like *Njallani* in cardamom have been developed by farmers in Idukki district. Kuttanad farmers have perfected the art and science of growing paddy below sea level. This knowledge will be of immense value in protecting coastal agriculture in the event of a rise in sea level.

Climate change presents mega-threats to Kerala's food and water security systems as well as to the lives and livelihoods of coastal communities. Sea level rise will cause serious threats to coastal ecosystems as well as to coastal mineral wealth, as, for example, the monozite and thorium deposits. A temperature rise of 2°C will affect the production and productivity of plantation crops like coffee, tea, spices and rubber, in addition to annual crops like rice. Change in precipitation may cause both drought and floods as well as soil erosion and decrease in soil fertility. The forest biodiversity and medicinal plant wealth of Kerala will also be adversely affected. Ecosystem services will be disrupted. Vector-borne diseases will affect plant, animal and human health. Kerala may experience a large influx of "climate refugees" from coastal to inland areas. To prepare for such threats, both anticipatory research using advanced technologies as well as participatory research with local communities including tribal families will be needed so that coping mechanisms combining frontier science and traditional wisdom can be developed and put in place soon.

Kerala has the unique advantage of becoming a world leader in managing the consequences of sea level rise. The Kuttanad area may be declared as a Special Agriculture Zone as it is the only region in India with experience of cultivating rice under below sea level conditions; it is a Ramsar site; it is a unique wetland promoting rice-fish rotation; it is an area of thriving water tourism; it is a biodiversity paradise in flora and fauna, and it provides uncommon opportunities for learning how to manage the impact of sea level rise. An International Research and Training Centre for Below Sea Level Farming should be established in Kuttanad.

The area extending from the Silent Valley Biosphere Reserve to Wayanad may be developed as a herbal bio-valley, to promote the conservation and sustainable and equitable use of the genetic diversity occurring in medicinal plants. Micro-enterprises supported by micro-credit may be organised by women's Self-Help Groups (SHGs) along the bio-valley. All Rare Endangered and Threatened (RET) species in the bio-valley should be protected and multiplied. The products of the bio-valley may be given a brand name. Conservation and commercialisation will then become mutually reinforcing, and there will be an economic stake in conservation. Today, there is an economic interest in the unsustainable exploitation of medicinal plant resources, and this needs to be halted and reversed through the medicinal plants biovalley.

Suggestions for the National Action Plan of the Biodiversity Authority of India

I would like to set out some of the steps we should take for making the conservation of biodiversity everybody's business.

Deliver as one

The importance of the use and conservation of biodiversity in agro-ecosystems should be recognised in national development plans (including poverty reduction programmes). This necessitates integration of approaches across government departments confronting rural development, food security, poverty reduction, environment and climate change. To the extent feasible the "deliver as one" approach should be adopted, in order to achieve convergence and synergy among different ongoing programmes.

Build partnerships

Effective use of agro-biodiversity is the key to realising its development impact and its conservation. This requires development of markets for products of diverse agriculture, especially underutilised crops, different animal genetic breeds, etc. This can be built on public–private partnerships and development of agribusinesses benefiting rural communities.

Strengthen the role of farming and tribal communities

Farming and tribal communities have a major role in delivering the benefits of agro-biodiversity. The following measures should be incorporated:

- Integrate community *in situ* and *ex situ* conservation in the national biodiversity conservation strategy. *In situ* conservation will start from the field. *Ex situ* conservation can take the form of sacred groves and heritage trees, as well as botanical and zoological gardens.

- Organise Field Gene Banks, Seed Banks and Grain Banks at the local level. This will help to promote *in situ* on-farm conservation of landraces and enlarge the food basket and thereby strengthen local level crop and food security.
- Establish special Gene Banks for climate-resilient crops.
- Recognise and reward primary conservers of biodiversity through initiatives like the Genome Saviour Award instituted by the Plant Variety Protection and Farmers' Rights Authority of India.

Conservation science

Research and development priorities should be re-focused to enhance the productivity of bio-diverse agriculture, including the need to optimise genetic diversity (plants and animals). For example, landraces and wild crop relatives should be characterised, evaluated and utilised in crop improvement programmes to transfer traits relevant to climate change, such as drought and heat resistance and flood and salinity tolerance. Breeding for per-day yield rather than per-crop productivity should receive priority. These concerns should be reflected in higher education curricula and research agenda.

Climate-resilient farming systems

Climate change will demand modifications to farming systems (e.g., cultivars, land use, water use management, animal selection) and increased environmental risk management. This will require prioritisation of the social and agro-ecological zones most at risk. Biodiversity is the feedstock not only for food and health security, but also for the management of climate change. Gene Banks for a warming planet have become urgent as an essential element of climate-resilient farming systems. The prospects for climate change have added urgency to efforts designed to save every gene and species now existing on our planet. The initiative of the Government of Norway in establishing a Global Seed Vault at Svalbard, and of the Defence Research and

Development Organisation of the Government of India in establishing a similar facility under perma-frost conditions at Chang La in the Himalayas are welcome steps.

Land-use patterns

Since land-use represents a third of global Greenhouse Gas emissions, this must be reversed. Farmers can help build soil carbon reserves and at the same time improve soil fertility through fertiliser trees and conservation agriculture. Mangrove forests are very efficient in carbon sequestration and should be protected.

Economic value of ecosystem services

According economic value to ecosystem services like land, water, biodiversity and climate and putting in place mechanisms for payment for such services will help to reduce the ecological footprint and thereby help to achieve a balance between biocapacity and natural resources exploitation.

Biodiversity literacy

An extensive and well-designed biodiversity awareness and literacy campaign should be launched, starting with schoolchildren and extending up to the adult population. Such a biodiversity literacy programme must involve the integrated use of traditional and new media. Village Knowledge Centres could be utilised for sensitising the local population on the threats to biodiversity and the names and locations of the RET species occurring in that area. University students and civil society organisations can be assisted in saving RET species. The preparation of local level Biodiversity Registers can be promoted. Genetic Gardens can be promoted in schools and colleges. There is need for promoting among the younger generation an awareness of the vital significance of agro-biodiversity for the well-being of the future generations. Genome Clubs designed to promote genetic and biodiversity literacy may be organised in all schools and colleges.

Climate care movement

The greatest casualty of climate change will be food and water security. Biodiversity helps to mitigate the adverse impact of climate change.

A climate care movement at the local, national and global levels must pay specific attention to the following:

- Gene care and conservation.
- Climate literacy.
- Appointment of local-level community climate risk Managers.
- Promotion of appropriate mitigation and adaptation measures.

In order to promote coordinated and concerted efforts in agro-biodiversity conservation and enhancement, it will be useful to constitute a Consortium for Agro-biodiversity Conservation and Enhancement in every State with members drawn from the government, academic, civil society, media and private sectors. Such a consortium should promote the conservation of germplasm of crops, farm animals, fisheries and forest trees. The consortium should help in ushering in an era of biohappiness arising from the sustainable use of bio-resources from creating more jobs and income.

To sum up, biodiversity is the prime mover of an evergreen revolution movement in agriculture, and the goal of achieving "**food for all and for ever**" is wholly dependent on its conservation and sustainable and equitable use. Mahatma Gandhi's plea that "we should live in harmony with nature and with each other" should guide community efforts in the management of biodiversity. The declaration by the United Nations that 2010 be observed as the International Year of Biodiversity is to remind humankind that biodiversity is the foundation for global food security and that conservation of biodiversity and natural resources should become a non-negotiable human ethic.

Chapter 3

Biotechnology and Biohappiness

The elucidation of the double-helix structure of the Deoxyribonucleic acid (DNA) molecule in 1953 by Drs. James Watson, Francis Crick, Maurice William and Franklin Rosalind marked the beginning of what is now known as *the new genetics.* Research during the last five decades and more in the fields of molecular genetics and recombinant DNA technology has opened up new opportunities in agriculture, medicine, industry and environment protection. The ability to move genes across sexual barriers has led to heightened interest in the conservation and sustainable and equitable use of biodiversity, since biodiversity is the feedstock for plant, animal and microbial breeding enterprises.

Considerable advances have been made during the last 25 years in taking advantage of the new genetics in the areas of medical research, production of vaccines, sero-diagnostics and pharmaceuticals for human and farm animal health care. The production of novel bioremediation agents as for example, the development of a new *Pseudomonas* strain for clearing oil spills in oceans, rivers and lakes is also receiving priority attention because of increasing environmental and water pollution.

There has also been substantial progress in agriculture, particularly in the area of crop improvement through the use of molecular marker-assisted breeding, functional genomics, and recombinant

DNA technology. A wide range of crop varieties containing novel genetic combinations are now being cultivated in USA, Canada, China, Argentina and several other countries. A strain of cotton containing the *Bacillus thuringiensis* gene (Bt cotton), which has resistance to boll worms, is now under cultivation in India based on both official and unofficial (illegal) releases.

There is little doubt that the new genetics has opened up uncommon opportunities for enhancing the productivity, profitability, sustainability and stability of major cropping systems. It has also created scope for developing crop varieties tolerant/resistant to biotic and abiotic stresses through an appropriate blend of Mendelian and molecular breeding techniques. It has led to the possibility of undertaking anticipatory breeding to meet potential changes in temperature, precipitation and sea level as a result of global warming. There are new opportunities for fostering pre-breeding and farmer-participatory breeding methods in order to continue the merits of genetic efficiency with genetic diversity.

While the benefits are clear, there are also many risks when we enter the territory of the unknown and unexplored. Such risks relate to potential harm to the environment and to human and animal health. There are also equity and ownership issues in relation to biotechnological processes and products. The following issues are the major areas of concern to the public and policy maker.

(a) *What is inherently wrong with the technology?*
Is the science itself safe, as for example, the use of selectable marker genes conferring antibiotic or herbicide resistance?

(b) *Who controls the technology?*
Will it be largely in the private sector? If the technology is largely in the hands of the private sector, the overriding motive behind the choice of research problems will be private profit and not necessarily public good. If this happens, "orphans will remain orphans" with reference to choice of research priorities. Crops being cultivated in rainfed, marginal and fragile environments, which are crying for scientific attention, may continue to remain neglected.

(c) *Who will have access to the products of biotechnology?*
If the products arising from recombinant DNA technology are all covered by Intellectual Property Rights (IPR), then the technology will result in social exclusion and will lead to a further enlargement of the rich–poor divide in villages.

(d) *What are the major biosafety issues?*
There are serious concerns about the short- and long-term impact of Genetically Modified Organisms (GMOs) on the environment, biodiversity and human and animal health.

There has been considerable debate in recent years on the potential impact of new biotechnologies on agriculture. The term biotechnology is currently being used to connote a wide variety of biological manipulations such as cell and tissue culture, embryo transplantation, transfer of DNA material across sexual barriers, vaccine production, bioremediation, microbiological enrichment of cellulosic material, fermentation and various forms of biomass utilisation. There are immediate opportunities for the multiplication of superior clones of fruit and forest tree species, as well as plantation crops like cardamom and oil palm through tissue culture methods. Enhancing biomass production and its conversion into energy are important applications.

The significance of biotechnology for a better biofuture of the Third World can be illustrated by taking the example of Asian agriculture. Asia has over 50 per cent of the global population, over 70 per cent of the world's farming families, but only 25 per cent of the world's arable land. At the beginning of the 21st century, the per capita land availability was 0.1 ha in China and 0.14 ha in India. The average Asian population growth rate is 1.86 per cent. The only pathway open to countries like China and India for feeding their growing human populations is continuous improvement in yield. This involves research which can further raise the yield ceiling. China has gone into the large-scale exploitation of hybrids in rice for this purpose. The tools of biotechnology can help in raising the productivity of major crops through an increase in total dry matter production which can then be partitioned in a way favourable to the economic part.

It is anticipated that soon over a quarter of worldwide seafood consumption is likely to be produced by aquaculture. Breeding programmes and genetic engineering have led to the production of new "boneless" breeds of trout that have a better feed conversion rate, and salmon which possess an antifreeze gene to enable them to survive in colder waters. Further research in fish breeding is expected to focus on growth acceleration, sex identification and determination, flesh quality, disease resistance, sea water adaptation, and the ability to utilise specific dietary components. Thus, biotechnological research is opening new windows of opportunity both in terrestrial and aquatic farming systems.

Biotechnology can be useful in a very different way too — in integrating brain and brawn in rural professions. For example, Kerala State in India is developing the district of Ernakulam as a Biotechnology District, for taking advantage of its rich educated human resource, particularly educated women, who often tend to be employed in unskilled jobs. The programme will include extensive tissue culture propagation of forest tree species, banana, cardamom and ornamental and medical plants, genetic improvement of cattle and poultry, and the establishment of biomass refineries. The cause of educated unemployment is often not the lack of employment opportunities *per se*, but the paucity of employable skills in educated youth. The prevailing mismatch between the skills needed for the sustainable conversion of natural endowments into economic wealth could be ended through a carefully planned learning revolution. Centres of training in biotechnology, based on the method of learning by doing, can play an important role in ending this mismatch.

In the coming decades, Indian farm women and men will have to produce more food and other agricultural commodities to meet home needs and to take advantage of export opportunities, under conditions of diminishing per capita availability of arable land and irrigation water and expanding abiotic and biotic stresses. The enlargement of the gene pool with which breeders work will be necessary to meet these challenges. Recombinant DNA technology provides breeders with a powerful tool for enlarging the genetic base of crop varieties and to pyramid genes for a wide range of economically important traits. The safe and responsible use of biotechnology will enlarge our capacity to meet the challenges ahead, including those caused by climate change. At the international level, the Cartagena Protocol on Biosafety provides a framework for risk assessment and aversion. At the national level, there is need for a regulatory mechanism, which inspires public, political and professional confidence.

Consumers all over the world are concerned with potential health risks associated with Genetically Modified (GM) foods. The nature and extent of concerns vary from country to country, depending upon the confidence the public have in the food and environmental safety regulation systems in place. For example, the Food and Drug Administration (FDA) of the United States attracts greater consumer confidence than the counterpart systems in Europe. The situation in India is similar to that in Europe. Public regard and satisfaction for the regulatory systems currently in place in the field of agricultural biotechnology are, to say the least, low. The bottom line for any biotechnology regulatory policy should be the safety of the environment, the well-being of farming families, the ecological and economic sustainability of farming systems, the health and nutrition security of consumers, safeguarding of home and external trade, and the biosecurity of our nation.

In contrast to GM crops, "life-saving" and "life-enhancing" GM pharmaceutical products seem to have more ready acceptance. The socio-ethical perspective often defines the public risk perceptions. Bio-ethical norms are as important as biosafety regulations in the case of medical and pharmaceutical biotechnology.

There are however some major concerns (elaborating the ones I have voiced earlier in this article) relating to current global trends in the objectives and organisation of technological research. First, the farm sector is a major export-earning enterprise for Third World countries. Therefore, there is genuine concern about the potential adverse impact of genetic engineering research directed at finding substitutes for natural products. Some examples are high-fructose corn sweetener as a substitute for sugarcane sugar, and substitutes for vanilla, cocoa and diosgenin extracted from the *Dioscorea* species.

A second major concern relates to the safety aspects of genetic engineering research. Will tests which are not permitted in the industrialised countries be done in the Third World? Will "super weeds" arise from research aimed at the development of pesticide and herbicide resistant crop varieties? Will the ecological ground rules underpinning the field testing of transgenic material be the same everywhere?

Third, the nutritive quality and food safety issues relating to genetically engineered strains and growth hormones need careful study, using criteria more relevant to conditions where undernutrition and malnutrition are widespread. Will crop varieties with multiple resistance to pests contain toxins which will ultimately affect the health of the human beings or animals consuming their economic parts? What kinds of safety evaluation procedures are needed for food ingredients produced by micro-organisms, single chemicals and simple chemical procedures and whole foods and other complex mixtures?

Fourth, will the biotechnology revolution help resource-poor farmers increase productivity largely with the help of farm-grown inputs? How can we design mutually reinforcing packages of technology, services and public policies which can ensure that all rural people — rich or poor, landowners or landless labour — derive economic and social benefit from new biotechnologies? Is it possible to design a pro-small farmer and pro-poor biotechnology programme?

Fifth, what will be the impact of the extension of IPR to individual genes and genotypes on the availability of such improved material to developing countries and resource-poor farm families? Also, will IPR be exclusively reserved for rewarding formal innovation, although the informal innovation system has played, and is playing, a key role in the identification and conservation of plant and animal genetic resources? What are the rights of the farm families who have conserved and selected genetic diversity, in contrast to the rights of the breeders who have used them to produce novel genetic combinations?

Sixth, will priorities in biotechnology research be solely market-driven or will they also take into consideration the larger interests and the long-term well-being of humankind, whether rich or poor? In other words, will orphans remain orphans in the choice of research priorities and investment decisions? For example, rice is the staple of nearly half the human population, most of whom live in Asia. Yet, the application of biotechnological know-how to solve some of the important problems in rice production would not have received the financial and scientific support they needed but for the decision of the Rockefeller Foundation to make a major long-term investment in this area. Human diseases like leprosy also illustrate this point.

Finally, there are environmentalists who believe that each technological fix to an ecological problem will ultimately generate new levels of ecological catastrophe. They therefore caution against venturing into unknown territories. This is particularly important in medical biotechnology, where issues of ethics frequently arise.

All these concerns can be met only by a proactive analysis of the potential beneficial and adverse impacts of biotechnological research, not only from the economic angle but also from the ecological, equity and ethical perspectives. Social scientists and ecologists should be involved in project design teams right from the beginning and should not just come at the end to make a post-mortem analysis. For biotechnology to lead to a better future for humankind, we need a systems approach, keeping in mind Albert Einstein's exhortation that the human well-being should be the ultimate objective of all scientific endeavour.

If we foster the safe and responsible use of biotechnology, we can initiate an era of biohappiness where biodiversity becomes the instrument of more food, income and jobs.

Chapter 4

Integrated Gene Management

On the eve of the UN Conference on Environment and Development held at Rio de Janeiro in June 1992, the Union of Concerned Scientists published an open letter titled *World Scientists' Warning to Humanity*, which stated that "human beings and the natural world are on a collision course." The letter stated further: "if not checked, many of our current practices put at serious risk the future that we wish for human society and the plant and animal kingdoms, and may so alter the living world that it will be unable to sustain life in the manner that we know". This warning was signed by over 1600 scientists from leading scientific academies in 70 countries. The list included 104 Nobel Laureates.

A Chinese proverb warns, "if you do not change direction, you will end up where you are headed". Since we do not want to reach where we are presently headed, what change of course should we bring about?

I wish to take biodiversity, one of the key components of our basic life support systems, as an example to illustrate what changes are needed in the management of our biological resources. It is now widely realised that the genes, species, ecosystems and traditional knowledge and wisdom that are being lost at an increasingly accelerated pace limit our options for adapting to local and global change, including potential changes in climate and sea level. It has been estimated that about 13 to 14 million species may exist on our planet. Of this, less than 2 million species have so far been scientifically described. Invertebrates and microorganisms are yet to be studied in detail. In particular, our knowledge of soil microorganisms is still poor. Also, biosystematics as a scientific discipline is tending to attract very few scholars among the younger generation.

Another important paradigm shift witnessed in recent decades in the area of management of natural resources is a change in the concept of "common heritage". In the past, atmosphere, oceans and biodiversity used to be referred to as the common heritage of humankind. However, recent global conventions have led to an alteration in this concept in legal terms. Biodiversity is now the sovereign property of the nation in whose political frontiers it occurs. Under the UN Convention on the Law of the Sea, nations with coastal areas have access to a 200-mile Exclusive Economic Zone (EEZ). For example, the ocean surface available to India under the EEZ provision is equal to two-thirds of the land surface available to the country. The Climate Convention and the Kyoto Protocol provide for both common and differentiated responsibilities to countries. Thus, the global commons can be managed in a sustainable and equitable manner only through committed individual and collective action among nations.

While we have some knowledge of variability at the ecosystem and species levels, our knowledge of intra-specific variability is poor, except in the case of plants of importance to human food and health security. The Global Biodiversity Assessment warns: "Unless actions are taken to protect biodiversity, we will lose forever the opportunity of reaping its full potential benefit to humankind". What kind of action will help us to ensure not only the conservation of biodiversity, but also its sustainable use?

The United Nations Environment Programme's Global Environment Outlook-2000 states:

> Reduced or degraded habitats threaten biodiversity at gene, species and ecosystems level, hampering the provision of key products and services. The widespread introduction of exotic species is a further major cause of biodiversity loss. Most of the threatened species are land-based, with more than half occurring in forests. Freshwater and marine habitats, especially coral reefs, are also very vulnerable.

How do we arrest this trend and prevent further genetic erosion?

I would like to summarise briefly the approach adopted in India as well as at the MS Swaminathan Research Foundation (MSSRF) to foster an integrated gene management strategy in the country. We use the term *management* in the context of natural resources to include conservation, sustainable use and equitable sharing of benefits. It is only such a concept of management that can result in the conservation as well as enhancement of natural resources.

The integrated gene management system includes *in situ*, *ex situ* and community conservation methods. The traditional *in situ* conservation measures comprising a national grid of national parks and protected areas are generally under the control of government environment, forest and wildlife departments. The exclusive control of such areas by government departments has often led to conflicts between forest dwellers and forest-dependent communities, and forest officials. The non-involvement of local communities in the past in the sustainable management of forests has resulted in a severe depletion of forest resources in India. It has become clear that sole government control alone will not be able to protect prime forests or regenerate degraded forests.

The Participatory Forest Management (PFM) procedure initiated first at Arabari in West Bengal in India in 1972, largely at the initiative of a young Forest Officer, Mr. Ajit Kumar Banerjee, and later extended to the other States in the country, became a significant turning point in the history of forest management in India as well as other Asian countries. The essential feature of this system is that the state and community become partners in the management of the forest resource. The state continues to own the resource but the benefits are shared. Access to non-timber forest products becomes an important avenue of sustainable livelihoods to the forest-dependent communities. Thus, the community develops an economic stake in the preservation of forests, leading to conservation and sustainable use becoming mutually-reinforcing components of a Forest Management Policy. The experience gained in India during the last few decades shows that the process of natural forest degradation can be reversed through PFM and that forests can provide the local community with non-wood forest products on a continuous or seasonal basis, if there is a more widespread understanding of their regenerating capacity. Since forests are home to a large proportion of naturally occurring biodiversity, saving forests results in saving genes.

We all live on this planet as guests of the green plants which help to convert sunlight into various agricultural commodities and of the farm families who toil in the sun and rain to cultivate them. There are millions of rural and tribal families who not only conserve the world's rich bioresources but also serve as the backbone of the global food and health security systems. Tribal and rural farming communities have a long tradition of serving as custodians of genetic wealth, particularly landraces often carrying rare and valuable genes for traits like resistance to biotic and abiotic stresses, adaptability and nutritional quality. Several land types that carry valuable genes are preserved by farmers for religious functions and they constitute valuable material for conservation and sustainable use. Local landraces are still being maintained largely by the tribal poor. They thus serve public good at personal cost. Such poverty-ridden custodians of genetic wealth are increasingly confronted with severe socio-economic problems which are rendering the maintenance of their traditional conservation ethics difficult.

Sacred groves are tracts of forest that have been completely immune from human interference on the basis of religious beliefs. Both *in situ* on-farm conservation of intra-specific variability, particularly in plants of food and medicinal value, and *ex situ* conservation through sacred groves have been part of the cultural traditions of rural and tribal families in India. In the Old Testament also, there are several references to sacred groves. Among the important trees usually preserved in Indian sacred groves are: *Ficus religiosa*, *Saraca asoca*, *Shorea robusta*, *Alstonia scholaris* and many other species of ecological, economic and spiritual value. Unfortunately, several of these traditions are now tending to wither away. It is only by giving explicit recognition to the pivotal role of community conservation in strengthening ecological, food and health security systems that we can succeed in the revitalisation of these traditions. In national integrated gene management systems, *in situ*, *ex situ* and community conservation methods should receive adequate and concurrent attention.

Commercialisation can become an important instrument of conservation, if commercial companies, such as pharmaceutical firms, will help rural and tribal families to cultivate rare medicinal plants on contract. This will help them to source the raw material they need, without directly exploiting plants growing in forest canopies. Many important medicinal plants are now in the Red Data Books of several developing countries due to the unsustainable exploitation of such plants from the wild. Domestication of economically valuable Red Data Book species will help the cause of conservation, while at the same time meeting the needs of the commercial user.

Methods of fostering the sustainable management of Biosphere Reserves were discussed at a conference convened by UNESCO in 1995 at Seville in Spain. The Seville vision states: "Rather than forming islands in a world increasingly affected by severe human impacts, Biosphere Reserves can become theaters for reconciling people and nature. They can bring the knowledge of the past to the needs of the future".

To convert the above vision into field level reality, we have been working for a shift in the nature of management of Biosphere Reserves in India. India has currently 85 National Parks covering 3.6 million sq km and 448 Wildlife Sanctuaries covering about 120,000 sq km in the major biogeographic zones. The total extent of protected areas includes 5 designated as World Heritage Sites, 15 Biosphere Reserves and 19 Ramsar Sites.

In 1994, MSSRF conducted a detailed study of the threats to the Gulf of Mannar Biosphere Reserve in Tamil Nadu. The study showed that unless the livelihood security of the families without assets living in that area can be strengthened, unsustainable exploitation of biological resources will continue. During 1996–1998, MSSRF initiated a proposal for bringing the Gulf of Mannar Biosphere Reserve under a participatory management mode.

This proposal was submitted by the Governments of Tamil Nadu and India to the Global Environment Facility (GEF) through the United Nations Development Programme (UNDP) for support. The GEF Council approved the proposal and commended it as a model project which deserves to be widely emulated by those preparing similar projects. The Scientific and Technical Advisory Panel (STAP) of GEF endorsed the project for approval with the following remarks.

> The project addresses a major challenge, namely the conservation of coastal biodiversity of the highest ecological value in a large area subject to considerable pressure from poor populations upon the sole resources that appear to be at their disposal. To meet this challenge, the project follows the only framework which can succeed, namely, to combine the necessary protection of the threatened ecosystem and ecological processes with economic and social benefits which will meet the essential needs of local people, through providing appropriate institutional, financial and managerial arrangements.

The management structure through which people and Nature will be united in the Gulf of Mannar area is through a Gulf of Mannar Biosphere Trust, comprising representatives of fishermen and rural communities as well as all the principal civil society stakeholders and concerned government departments. It is hoped that the Gulf of Mannar Biosphere Trust, whose long term sustainability will be ensured with the help of a Trust Fund, will show the way for promoting a management by a partnership system of governance in all the other Biosphere Reserves in the country.

A major need in such programmes is the strengthening of the livelihoods of the poor families living in the vicinity of the Biosphere Reserve. For this purpose, the biovillage model of livelihood security will be introduced in the villages around the Reserve. The biovillage concept of human-centered rural development aims to address concurrently the challenges of natural resources conservation and poverty eradication. Market-driven livelihood opportunities will be identified and local families will be assisted in taking to them with the help of institutional credit.

It is now widely recognised that the conservation continuum consists of three links.

In situ	→	On-farm	→	*Ex situ*
Conservation of habitats		On-farm conservation by rural and tribal communities		Conservation through botanical and zoological gardens and cryogenic gene banks

While the two ends of this conservation chain (*in situ* and *ex situ*) receive support from public funds, on-farm conservation by rural and tribal women and men largely remains unrecognised and unrewarded. Yet, this link in the chain is responsible for the conservation of valuable intra-specific variability.

MSSRF's partnership with local communities and government agencies is designed to strengthen this neglected component in the conservation chain. For this purpose, a Community Gene Management System (CGMS) has been developed, comprising

- Field Gene Banks (FGBs) at the village level, with *in situ* on-farm conservation by rural and tribal families through participatory selection and breeding.
- Area Seed Banks (ASBs) at the level of a cluster of villages, where the conservation of representative samples of seeds in seed stores will be the insurance against total loss of seeds during drought years.
- Community Gene Banks (CGB) and Herbariums, where cryogenic preservation will be useful as evidence for getting tribal families reward and recognition under the Plant Variety Protection and Farmers' Right Act.

FGBs are basically *in situ* on-farm centres of conservation. Landraces and location-specific Plant Genetic Resources (PGR), usually identified on a participatory basis with the local families, are conserved in FGBs. The local landraces are periodically grown in their native habitats for seed renewal. They serve as effectively decentralised and highly cost-effective arms of a CGMS. No expensive infrastructure is needed to set up FGBs. The communities in natural resource areas would be stakeholders in the FGBs. They can then provide active support not only to prevent genetic erosion in those areas, but can add new genetic stocks as and when they discover them. FGBs serve the cause of both conservation and continuous evolution through selection, recombination and mutation.

Several FGBs can be linked to an ASB, taking into account factors like distance, communicability, conservation space and the like. There could be two to three seed banks in a district. The ASB, managed by a committee of the stakeholders, will help to strengthen coping mechanisms for facing the problem of seed scarcity caused by drought-induced crop failures. Logically this committee can monitor Intellectual Property Rights (IPR) issues such as prior informed consent and benefit sharing.

One or more ASBs will be linked to the CGB, which will hold *ex situ* the seed stocks of landraces, etc., along with herbarium sheets and other information needed to get the primary conservers reward and recognition under the Indian Act for Plant Variety Protection and Farmers' Rights. The CGB could deposit duplicate sets of accessions in the National Gene Bank (NGB) of the National Bureau of Plant Genetic Resources at New Delhi.

The CGMS System provides an opportunity for fostering symbiotic partnerships between rural/tribal women and men and scientists in areas like participatory breeding and the development of new varieties adapted to local conditions using novel genetic combinations provided by genetic enhancement centres. In addition to the revitalisation of the traditional conservation ethic of local communities, such a system will help to implement the conservation and equitable sharing provisions of the Convention on Biological Diversity (CBD).

Genetic homogeneity associated with mono-cropping and modern agriculture leads to the replacement of large numbers of local varieties with a few high-yielding strains. The dangerous consequences of covering large contiguous areas with one or two genetic strains are now well known. Varietal diversification and crop rotations involving crops with non-overlapping pest sensitivity are important for sustainable agriculture. The transition from "green revolution" to an "evergreen revolution" involves the substitution of a commodity-centered approach with a farming-systems approach. The farming-systems approach involves the adoption of mixed farming (crop-livestock-fish) methodologies, based on an integrated natural resources conservation strategy. Participatory breeding including the use of novel genetic combinations arising from genomics and molecular breeding will help to combine advances in yield and quality with location-specific adaptation.

With the growing privatisation of plant breeding and expansion of proprietary science, it is important that an ecologically, economically and socially sustainable farming systems policy is developed for each agro-ecosystem. Such a policy will have to be developed jointly by farming families, official extension agencies and private sector companies. Unilateral introduction in large areas of one or two genetically modified strains of important food crops could cause irreparable harm in a few years' time both to food security and technological credibility. The pathway to an evergreen revolution on the farm is the adoption of integrated natural resource and gene management strategies.

In most developing countries, particularly in India and China which have 50 per cent of the global farming population, continuous advances in farm productivity per units of land, water and energy are essential for sustainable food security. Hence, there is need for developing and disseminating ecotechnologies, based on appropriate blends of traditional wisdom and technologies with bio-information, space and renewable energy technologies. The ecological prudence of the past and the fruits of contemporary innovation can then be combined in a symbiotic manner.

It is important that in all the features of a community-centered integrated gene management strategy, gender roles in all aspects of genetic resources conservation, sustainable use and equitable sharing of benefits are given attention. Women in many developing countries are the primary seed selectors and savers. Their contributions to the evolution of a biodiversity conservation ethic should be fully recognised in any system which is designed to operationalise the equity in benefit-sharing provisions of CBD.

If we promote worldwide a community-centered integrated gene management strategy, we will soon stop hearing about vanishing species and vanishing landscapes and habitats. None of the policies and procedures I have suggested are difficult to implement. What is needed is as much interest in preventing species extinction and biodiversity loss, as in the chronicling of threats and preparation of *Red Data Books*.

Today, governments and the scientific community often play the dominant role in organising and managing structured *in situ* and *ex situ* conservation measures. By recognising that conservation efforts represents a continuum, with rural and tribal families performing a vital function in preserving precious genetic variability in important plants and farm animals, we will be able to attend to all the links in the conservation chain. Community conservation is a value-added link in the conservation system, since local families not only conserve but add value to the conserved material through selection and information.

Let me conclude with the hope that if the 20th century was a period of understanding and chronicling threats to biodiversity and bioresources both in land and water, the 21st century will be one where the threats are terminated and benefits harnessed for a better common present and future for humankind.

Chapter 5

Mangroves

The coastal zone is characterised by a large variety of forms which have been shaped in the course of geological history. These are rocky coasts, beaches, river mouths, fjords, estuaries and lagoons, barrier islands, intertidal flats and wetlands. In and around them, a number of specific biological communities have developed, including intertidal and marsh communities, mangroves, sea grass beds and coral reefs. Together, coastal systems represent an almost infinite variety of riches in which creatures have abundantly developed, both in species and numbers. Man too has settled in the coastal zone for partly similar reasons such as easy accessibility, abundance of food and protection against enemies.

A diverse group of predominantly tropical trees and shrubs growing in the marine intertidal group — where conditions are usually harsh, restrictive and dynamic — is called mangroves. The main tree flora of mangroves consists of highly specialised plant species, well-adapted to the unique conditions of seawater inundation during high tides throughout the year and nearly fresh water conditions during monsoon, being constantly subjected to both rhythms of tides and seasons (shorter term) as well as changes of climate and sea level (longer term). The interesting characteristic of the system is that only a few genera and specific species with convergent adaptation to the group of selection forces are present all over the tropical and subtropical estuarine tidal regions of the world. This living together as a community subject to continuous changes like slicing, colonising and erosion in the seaward side or change of rise in the tides and drainage on the landward side is an intriguing feature.

Other important features are the aerial roots, vivipary and new colonies and incipient islands in shallow water which may survive or become extinct over a period of time. The composition of the vegetation is low to medium in tall tress, shrubs and salt marsh herbs. Although mangroves are able to grow on sand, peat and coral, the most extensive mangroves are invariably associated with mud and muddy soils. Such soils are usually found along delta coasts, in lagoons, and along estuarine shorelines. The habitat in which mangroves are observed is often referred to as mangrove forest and/or tidal forest.

The origin and distribution of mangroves is well documented. The geological history and evidence show that mangroves appeared between the Eocene and Oligocene periods (30 to 40 million years ago). Plant remains or fossils of major mangrove genera like *Rhizophora, Nypa* and others provide important landmarks. The oldest among these species is *Nypa* which is supposed to have originated during the end of the Cretaceous period, while species like *Pelliciera, Rhizophora* are from the Eocene period which perhaps makes them some of the oldest members.

The diversity in a mangrove system has two components, namely: (a) species diversity and (b) genetic diversity within species. These two components are closely intertwined and inseparable. For example, *Avicennia* is present on the firmer, exposed seaward side, while *Sonneratia* is in soft, rich mud along sheltered river mouths and *Ceriops decandra* grows in the sheltered coast. However, *Bruguiera* exists in varied conditions and the same is true for *Rhizophora. Rhizophora mucronata* is abundant in sandy, firmer bottoms and also regions of low tide areas bordering the vegetation, while *R. apiculata* is found in the banks of tidal coasts.

Mangroves are mostly out-breeders, sexually propagated and with diversity of floral biology, pollination and breeding mechanisms. Growth in such an environment, with diurnal and seasonal fluctuations in physical variables, and the patterns of convergent adaptation, make mangroves an ideal material for geneticists both from evolutionary and conservation points of view, and for formulating policy for monitoring and controlling changes.

Mangrove forests serve as links between terrestrial and marine ecosystems. Import of nutrients from the land to the mangroves and export of organic matter from mangroves to the sea is one of the important features of energy flow. The litter fall of mangrove trees amounts to a large quantity of nutrients which are circulated both within the mangrove forests and in the surrounding marine ecosystem. Accumulation of rich nutrients amounts to the increasing productivity of the bottom sediments of mangrove swamps. The forests also serve as nursery grounds for a variety of fishes and prawns. This is why productive fishing grounds such as those of the rivers Ganga (India) and the Rung Sat (South Vietnam) are adjacent to the mangrove swamps.

Mangrove Genetic Resources of India

In India, the important mangrove forests are found in the Andaman and Nicobar Islands as well as the Sunderbans on the eastern side and also throughout the eastern and western coasts of the mainland. The mangroves occur in various habitats such as delta, estuarine, backwater and sheltered-insular bay types. The great rivers such as the Ganga, the Brahmaputra, the Mahanadi, the Godavari and the Krishna discharge enormous quantities of fresh water along with silt, particularly during the flood season. The vast delta mangroves of the Bay of Bengal as well as those along the delta coasts of Odisha, Andhra Pradesh and Tamil Nadu owe their luxuriance to this unique natural process of habitat formation. Bhitarkanika in Odisha is particularly rich in mangrove genetic diversity. This has resulted in species richness of the mangrove forests and greater forest cover on the east coast (about 70 per cent of the total mangrove forest area in India) as compared to the western coast.

Out of about 180 species of exclusive and non-exclusive species of mangroves that are observed throughout the world, over 50 per cent are represented in India. The Indian mangroves (about 97 species) belong to 62 genera covering over 37 families.

Flora species which can sustain various levels of salinity in the elevated (sub-aerial) regions, above the spring tidemark, are termed as non-exclusive mangrove species or back mangroves. Over 30 terrestrial species are known to grow in these upper shore areas. Sea grasses, seaweeds, microbial flora such as yeasts, bacteria and fungi, phytoplankton species and such like are also part of the mangrove ecosystem.

Zooplankton, wood borers, fishes, shell fishes and crustaceans constitute the smaller fauna in the mangrove forests of India, besides a great deal of wildlife diversity. The Royal Bengal tiger is one of the unique resident species of the mangroves of the Sunderbans. Monitor lizards, estuarine crocodiles, as well as various species of monkeys, otters, deer, fishing cats and wild pigs are some of the most common species found here. The most common species of resident and migratory birds found in the mangrove forests include kingfishers, herons, storks, sea eagles, kites, sand pipers, curlews, terns, etc. Flamingos flock to the exposed mudfiats during low tide.

Diversity of life in marine coastal ecosystems was not understood till recently and marine conservation has only become an issue of global concern within the last 20 years or so. Far less publicised than the loss of biodiversity on land, the loss of marine coastal genetic, species and ecosystem diversity is a global crisis.

A principal cause of coastal ecosystem degradation is the unsustainable exploitation of natural resources stemming from the rapid population growth. A second set of problems pertains to the pollutants released in the marine environment. Mangroves and coral reefs are the most threatened ecosystems. They provide a substantial part of the protein intake. Reefs and non-reef coral communities within 15 km of shore are generally overfished, while major destructive forces include sedimentation related to deforestation and various forms of destructive fishery, diversion of mangrove areas to other uses, e.g., shrimp farming, offshore tin mining and dumping of hazardous and unwanted material.

Certain natural phenomena constitute a threat to mangroves. For instance, storms and hurricanes have often been responsible for the destruction of mangrove vegetation. In addition to the primary effects such as stripping of leaves, breakage of small branches and uprooting of trees, secondary effects have also now and then contributed to their destruction. However, the human hand has come down far more heavily on this vegetation than the occasional natural phenomena. Owing to their location close to the sea, these areas have also attracted alteration for the expansion of housing areas, roads, ports, industrial complexes, mining and man-made recreation centres, and so they have been subject to considerable development pressures.

The Inter-governmental Panel on Climate Change, in its scientific assessment of the prospects of sea level rise, concluded that for the business-as-usual scenario in year 2030, the global mean sea level could be 8 to 29 cm higher than its current level, with a best estimate of 18 cm, while in year 2070 it could be 21 to 71 cm with a best estimate of 44 cm. Most of the contribution is estimated to derive from thermal expansion of the oceans and the increased melting of mountain glaciers. The rise of the mean sea level will thus have significance for the future of not only island nations but also of the tropical tidal ecosystems. It will have an immediate and direct effect on ecosystems of the intertidal zone, with decline in the influence of terrestrial processes at all locations, and increase in the influence of marine processes. Disruption of established intertidal forests and relocation on formerly more terrestrial surfaces as these become inundated by the rising sea may not be so simple, with alteration of ecological and sedimental processes.

In view of the considerable cost of human activities in the normal processes of climate and radiation regulation and to meet different climatic possibilities and ocean levels, the MS Swaminathan Research Foundation (MSSRF) initiated in 1990 an anticipatory research through the project entitled "Genetic engineering and adaptation to climatic changes: Establishment of a genetic resources centre for identifying and conservation of candidate genes for use in the development of transgenic plants" in the Pichavaram Mangrove Forest with financial assistance from the Department of Biotechnology, Government of India. The highlights of the work are described below.

Conservation

With a view to preventing further destruction of mangrove forests, for sustained improvement and utilisation of mangrove forest genetic resources, and to conserving and enhancing the biological diversity in mangrove ecosystems, an integrated approach for the preparation of a "global strategy" is required. The conservation and preservation of mangrove genetic resources is the first step towards this end.

The Foundation pioneered efforts in preventing further erosion of the existing mangrove germplasm wealth and developing an integrated approach to *in situ* and *ex situ* conservation and sustainable utilisation of mangrove forest genetic resources. Based on the criteria developed by an inter-disciplinary team of experts, the site Peria Guda was identified for *in situ* conservation and for the establishment of a Mangrove Genetic Resources Conservation Centre (MGRCC). This core area of 50 hectares was kindly made available on lease for 40 years by the Department of Forests, Government of Tamil Nadu. In order to conserve a representative sample of genetic diversity in mangrove species, an *ex situ* conservation site was identified in Peria Guda. Consolidation of genetically important material of mangroves from different parts of the country has been done in this area.

Realising the importance of conserving a representative sample of genetic diversity in all the mangrove species in India, the need to establish link centres of MGRCC at the national level was felt. Three potential sites — Koringa (Andhra Pradesh), Bhitarkanika (Odisha) and Chorao (Goa) — were evaluated, based on the suggestion from the Ministry of Environment and Forests, Government of India. These link centres will serve as genetic gardens to conserve and enhance genetic diversity in mangroves.

Restoration

The mangrove ecosystem, though open, is quite complex, being composed of various inter-related elements in the land–sea interphase zone. Mangroves are known to keep the shoreline intact against tidal currents by preventing soil erosion. In view of the ecological and socio-economic importance of these plants, their restoration has become increasingly important, especially in recent years when the land cover of the earth is rapidly on the wane.

Development of suitable forestry methods for propagation and artificial regeneration of mangroves have been prompted by the demand for economically valuable mangroves; however, not much has been achieved in this respect. The work on experimental plantation of mangroves was based on: (a) nursery experiments (germination and growth of mangrove species), and (b) pilot plantations of mangroves. Based on this experience, large-scale plantations were undertaken which confirmed the development of location-specific plantation techniques for eco-restoration. Later, a team of MSSRF scientists standardised techniques for restoration of the upper shore areas in the intertidal regions of the mangrove forest.

Tissue culture regeneration studies were undertaken in various mangroves species. These efforts gave promising results and species-specific protocols were developed for as much as three species of mangroves for the first time anywhere.

Evaluation

Mangroves are distributed according to three important scales, namely their coastal range, location within an estuary and their position along the intertidal profile. The distribution pattern of mangroves in an estuarine region depends on several factors. Influence of freshwater run-off and estuary size is always felt when composition of mangrove flora is observed. In larger estuaries, there is a greater range of specialised habitats, and hence the presence of more species as compared to those in smaller estuaries. There is a general trend between genera and distribution, in such a way that genera with the greatest number of species consistently occur in a greater number of biogeographic regions. In addition, there is also isolation at other scales of distribution, notably in the specialisation of particular mangroves for certain habitats. Conventional genetic studies are difficult in mangroves and many forest tree species. In view of the difficulties and delay in conventional genetic analysis, standardised molecular methods of genetic analysis using the molecular variation in Deoxyribonucleic acid (DNA) was one of the approaches adopted to meet this challenge. In Phase II (1993–1995) of this project, work on molecular genetic studies was intensified.

Classification

It is common these days to come across ill-defined morphological characters of tropical plants. Mangroves are no exception to this. Considering the limited number of these unique plants and the extent of the variation of species level, it is important to sort out these constraints and therefore, it would be useful to identify genetic differences so as to remove the doubts and subjectivity surrounding the diagnostic character in the systematics of these paints. Morphological classification system was developed for the world's mangroves and their associates. The species were described based on morphological characters, and within the species variations were identified based on extensive field surveys. This was supplemented with the creation of lined drawings for selective species.

Under a project sponsored by the International Tropical Timber Organisation (ITTO), Japan, a Mangrove Ecosystems Information Services (MEIS) was commissioned. This forms the world's largest database on mangrove ecosystems. The project under the auspices of the Department of Biotechnology marked the beginning for providing a base to conceptually design such databases.

Education

Human and mangrove interaction has been evident from the studies undertaken in various parts of the world. The areas with mangrove ecosystems are always under stress as the resources are used by the local population for their livelihood security. For example, the bulk of mangrove resources has been used for firewood, a marginal livelihood activity. In this context, the population residing in the vicinity of such mangrove areas needs to be made aware of how mangroves are important and why they should be conserved. Therefore, a community-based approach needs to be brought about for restricting over-exploitation of mangrove forests. At the same time, a detailed analysis of the social behaviour of coastal communities can achieve a hope of success in this approach.

A Sustainable Livelihood Security Index (SLSI) developed by MSSRF outlined a community-based approach for ecologically sustainable utilisation of mangroves. Extensive surveys were carried out in various coastal regions of India which helped to understand the indigenous knowledge base and provided a basis for suggesting ecologically sustainable management practices to the coastal communities.

The entire project was executed by MSSRF in collaboration with the Forest Department, Government of Tamil Nadu. The research activities were directed by the Steering Committee constituted by the Department of Biotechnology, Government of India, which has funded the project. Besides, national and international teams extended expert guidance in its execution.

Where Do We Go From Here?

As in most tropical regions, mangrove swamps in India is a seriously endangered ecosystem. The main threat comes from over-exploitation, i.e., clear-cutting and land reclamation. In certain areas, pollution induced by the human factor also causes serious destruction. While considering the rapid growth of human population together with the key role of mangroves in the preservation of coastal and offshore fish and shelf-fish stocks and the wave-breaking, shore-protecting properties of the ecosystem, the protection and management of mangrove swamps is highly essential. It is also a fact that the major part of the population using the mangroves environment for their income is the rural poor in all countries. Any change will, therefore, affect them first. Thus, for example, the extensive and irrational destruction of the once huge mangrove swamps along the coasts of South East Asia, notably Bangladesh, is believed to be the main cause for the repeated catastrophic flooding in these areas in the last decade. Such extreme calamities can be averted by conservation of mangrove communities which are the barriers between the mighty water masses and land.

The importance of tropical rainforests as rich reservoirs of biological diversity is widely recognised. Mangrove forests in coastal belts of tropical and subtropical regions are equally important natural reservoirs of biological diversity. Moreover, the mangrove ecosystem constitutes a bridge between terrestrial and aquatic ecosystems and provides numerous benefits to coastal populations. These delicate ecosystems are seriously threatened by human interference. For sustained improvement and utilisation of mangrove forest resources, it is essential to conserve the existing mangrove species and genetic diversity within them. Like many agricultural and forest species, an integral global strategy is required to conserve and enhance biological diversity in mangrove ecosystems. The understanding of genetic diversity among mangroves, thereby identifying criteria and evaluation methods for forest areas for the establishment of MGRCCs, has been the first step in this direction.

Chapter 6

Plant Variety Protection and Genetic Conservation

The free movement of germplasm across regional, national and international boundaries is a fundamentally important process in Plant Genetic Resources (PGR) utilisation. Germplasm of domesticated and semi-domesticated plant species has moved freely with human movement throughout the history of agriculture. This migration, accompanied by adaptive divergence, has produced new diversity within and among species, and enriched the global biological wealth. Thus, the overall result of these processes in crop evolution has been beneficial to the conservation of biological diversity. Apart from national quarantine regulations, there had been little restriction on the movement of germplasm across political boundaries. Plant explorers and collectors, including the celebrated NI Vavilov, had enjoyed free access to germplasm from both within and outside their own countries. Indeed, the bulk of the progress in modern plant breeding in developed countries was based on the unimproved but valuable landraces and other germplasm materials introduced and collected from centres of genetic diversity occurring in developing countries. In the past few decades, however, the spatial movement of plant germplasm had been subjected to quarantine regulations as a preventive measure against the introduction of new pests and diseases.

Much of the spectacular successes in plant variety development in the rich industrialised countries in the North are attributed to the richness of genetic diversity at the centres of origin and primary diversity of economic species, located in the poorer developing countries of the South. While the genetic indebtedness of the North to the South is widely recognised, sharing the economic benefits accruing from genetic wealth is still a matter of considerable debate and discussion. Unfortunately, however, this debate is complicated by the enactment of plant variety protection legislations, commonly known as the Plant Breeders' Rights (PBR), in most of the industrialised countries. The presumed goals of PBR, namely (a) rewarding creativity of successful plant breeders, and (b) stimulating plant breeding activities in the private sector, are consistent with the free-market philosophy of the Western world. Because the PBR laws are initiated and enacted mainly in developed countries, these have created considerable international mistrust and pose a threat to the free flow of genetic resources. The majority of plant breeders and conservation specialists in the North tend to believe that the presumed linkage between PBR and Plant Genetic Resources (PGR) is imaginary, or tenuous at best. On the other hand, most breeders and scientists in the South find a strong linkage, and passionately feel that, like the intended goal of the PBR to reward and encourage modern plant breeders, the early contributors of genetic diversity to modern plant breeding deserve both moral recognition and material reward. They appear apprehensive of the double standards inherent in the argument for universal sharing of the world's genetic heritage, but restricting financial rewards to the breeders in the North. Thus, instead of promoting the spirit of sharing and international co-operation, the PBR issue has helped to polarise the gene-donor and -recipient countries into opposing camps. The overall negative impact of this polarisation on developing a comprehensive global strategy for genetic conservation cannot be overstated.

As mentioned earlier, the primary genetic resources used in modern plant breeding in the North originated in the South, where PBR is viewed as unfair, for the reward systems recognise only the last actor(s) in a long chain of biological advance. The situation can be illustrated from the pedigree of a highly successful modern variety of rice — IR 66 — which reveals that over 20 landraces of rice have contributed key genetic materials conferring resistance to several important biotic and abiotic stresses. These donor landraces are the products of selection and multiplication by rural women and men in several countries in the South over long periods of time. It has been argued, and with good reasons, that the development of IR 66 cannot be viewed as an isolated event, but as a series of historical events, credited to an unknown number of pre-breeders. A vast majority of the pre-breeders preserved the old and traditional agro-ecosystems, which domesticated and enriched most cultivated species of economic importance. The need for recognising such informal innovation led to the development of the concept "Farmers' Rights" in the Food and Agricultural Organisation (FAO) forums.

Considerations like the above led the Keystone International Dialogue Group to develop the concept of GIFTS in dealing with matters relating to PGR. The alphabets of GIFTS stand for the following:

G: Germplasm
I: Information on genetic make-up
F: Funds
T: Technology, including genetic engineering
S: Systems, including diverse cropping and farming systems.

Under such a concept, all nations are both donors and recipients in one or the other of the GIFTS package.

Since profit provides the only incentive for private commercial organisations to consider involvement in the business of plant breeding, PBR is an essential prerequisite for attracting the private-sector investment in this business. This is why in the global Biodiversity Convention signed by many nations at Rio de Janeiro in June 1992, there is emphasis on respect for Intellectual Property Rights (IPR). However, the question will always remain: How far should contemporary society go in privatising a public resource of inestimable value to global food security?

Although it is intuitively clear that the commercialisation of plant breeding would hamper the free flow of germplasm, some experts, particularly in developed countries, believe that there is no apparent reason for reduced exchange of PGR, both within and among countries, and between the North and the South, on account of PBR. While some experts have found no evidence of PBR impeding the free exchange of germplasm, there are indications that interest in private sector plant breeding in the USA is attributable, at least in part, to the American PBR laws. Justified by the increased private-sector involvement in plant breeding, the rate of decline of public plant breeding is likely to be accelerated by the shrinking public-sector (e.g., government) support. What are the implications of such a trend?

Consider, for example, the potential conflict between short-term financial benefits and long-term biological gains. A private plant breeder under highly competitive free-market conditions would be most concerned with the next plant variety, and least concerned about genetic conservation for the next millennium, or even the next century. It does not make a lot of economic sense to place a higher priority on long-term genetic safety than on short-term commercial success. As the goals of commercial plant breeding and genetic conservation are different, the *modus operandi* of the two operations may be conflicting. In that instance, the choices of private plant breeding programmes are obviously limited.

The profit incentive of commercial plant breeding naturally discourages investment in research on minor economic species. As profit is directly related to the volume of trade, natural targets of private plant breeding are the major crops with greater promise of profits. Furthermore, there would be the natural inclination to develop cultivars with wide, rather than specific, adaptation. Whereas wide adaptation of plant varieties makes sound economic sense, ecological sustainability demands diversity and location-specific varieties.

The advent of the era of molecular biology and recombinant Deoxyribonucleic acid (DNA) research has brought home the point that all forms of genetic diversity have potential commercial value and therefore need protection. As I keep reiterating, the basic feedstock for the biotechnology industry is biodiversity. This is why in the global Biodiversity Convention the linkage between the two has been stressed. In the GATT–TRIPS negotiations there have been sharp differences on the appropriateness of patenting living organisms. On the other hand, there are indications that the scientists involved in the sequencing of DNA in the human genome will apply for patents for specific base pair sequences. The same may happen in the case of the rice genome and *Arabidopsis* genome projects. There is little doubt that if such trends continue unabated, international collaboration in breeding research as well as in saving and sharing of biological diversity will suffer a serious setback. This should not be allowed to happen at least in crops and farm animals of importance to global food security.

This is an issue which deserves priority attention, among other concerns, such as methods of recognising and rewarding informal innovations represented by the efforts of rural women and men in the conservation and improvement of landraces and local cultivars; management and utilisation of existing *ex situ* collections of PGR in developed countries and in international agricultural research centres; and achieving a balance between public good and private profit in crops fundamental to national and global food security.

The progress witnessed during recent decades in keeping the rate of food production above the rate of population growth is largely due to genuine partnership among scientists belonging to many disciplines, institutions and countries. Such partnership has enriched science and insulated society from food famines. We may still be able to save humankind from Malthusian predictions if the following three objectives are steadfastly kept in mind:

1. Contain and prevent genetic erosion worldwide.
2. Conserve and utilise genetic resources effectively.
3. Share the benefits equitably.

Section II

Science and Sustainable Food Security

Chapter 7

The Tsunami and a New Chapter

The tsunami of 26 December 2004 was a terrible calamity resulting in serious loss of lives and livelihoods in the coastal areas of Tamil Nadu, Andhra Pradesh, Kerala, and the Andaman and Nicobar Islands. The response to this calamity was immediate from the Central and State governments, non-governmental organisations, bilateral and multilateral agencies, UN organisations, religious groups, and the media. In an article titled, "Beyond Tsunami: An Agenda for Action" in *The Hindu* of 17 January 2005, I outlined the immediate as well as the short- and long-term measures that should be taken for providing relief to the affected families, and for strengthening the coping capacity of the coastal communities in case of future tsunamis. I also indicated the steps needed to strengthen the ecological security of coastal areas, in order to ensure sustainable livelihoods for both the fishing and farm communities living along the coast. This agenda for action served as the basis for the tsunami recovery plans of many government and non-government organisations.

The tsunami served as a wake-up call for both government and community management of our coastal areas. Nearly 250 million people live within 50 km of the shoreline, in addition to about 5 million fisherfolk. The fishing communities are, unsurprisingly, the most affected during tsunamis, cyclonic storms, floods and tidal surges. Fisher families live on the coast but depend upon the sea for their livelihood. The tsunami underlined the need for an integrated approach to the management of the coastal zone.

The length of India's coastline ranges from 1962 km in the Andaman and Nicobar Islands, 1600 km in Gujarat, and 1076 km in Tamil Nadu, to 142 km in the Lakshadweep Islands. From 1991, the management of the coastal zone has been regulated through the Coastal Regulation Zone (CRZ) Notification under the Environment (Protection) Act, 1986. A Committee constituted by the Ministry of Environment and Forests in 2004 under my chairmanship examined the operational difficulties experienced in implementing the CRZ Notification.

We recommended that, instead of regulating only the use of the landward side of the sea, we should take both the sea and land surface for sustainable and equitable management. The inclusion of the sea surface is important to prevent pollution, erosion, and salt water intrusion as well as for facing the challenge of sea level rise caused by global warming and climate change. In a draft notification issued for public debate by the Union Ministry of Environment and Forests on 21 July 2008, the coastal zone has been defined as "the area from the territorial waters limit (12 nautical miles measured from the appropriate baseline) including its sea bed, the adjacent land area along the coast and inland water bodies influenced by tidal action including its bed, up to the landward boundary of the local self government or local authority abutting the sea coast, provided that in case of ecologically and culturally sensitive areas, the entire biological or physical boundary of the area may be included as specified under the provisions of Environment Protection Act, 1986".

The bottom line of any integrated coastal zone management strategy should be safeguarding the ecological security of coastal areas, the avoidance of sea pollution as well as unsustainable exploitation of living and non-living aquatic resources, protecting the livelihood security of fishing and farming communities, and the conservation of cultural heritage sites as well as migratory routes of birds and the Olive Ridley turtle and other faunal breeding grounds.

The fisher families, whose only source of livelihood is living aquatic resources, are concerned that the draft CRZ Notification of 2008, if implemented, will open the doors to depriving them of their housing sites and access to the ocean, because of the land grab tendencies among the rich. These are genuine concerns based on past experience. Therefore it will be desirable to enact legislation along the lines of the Scheduled Tribes and other Traditional Forest Dwellers (Recognition of Forest Rights) Act 2006 to safeguard the interests and rights of the fishing communities. This will ensure the long-term security of the sole means of survival for more than five million fishermen and women living near the sea.

In future, the greatest threat to coastal communities will come from a rise in sea level as a result of global warming. The President of Maldives, for example, has been highlighting the threat to the survival of his nation posed by a rise in sea level. We will face similar threats to the Andaman and Nicobar Islands, the Lakshadweep group of islands, and the coastal areas in the mainland, including cities like Chennai, Mumbai, and Kolkata. Recently, the Government of India launched a National Action Plan for Climate Change comprising eight Missions. Although a reference is made in the Plan to take proactive action for preventing a serious loss of lives and livelihoods when the sea level increases in areas adjoining the oceans, it will be prudent to have a separate Mission for managing the consequences of sea level rise, because this will decide the future of nearly 250 million children, women, and men.

The mangrove and non-mangrove bioshields I recommended in *The Hindu* article of January 2005 have now become part of the National Disaster Management Plan. Because of the outpouring of support for post-tsunami rehabilitation from many donor agencies, non-governmental organisations could undertake several useful long-term measures. For example, scientists of the MS Swaminathan Research Foundation (MSSRF) have undertaken the restoration, rehabilitation, and creation of bioshields in Tamil Nadu and Andhra Pradesh, which will serve as effective speed breakers when a tsunami-like situation arises in the future. Over 200 hectares of bioshields have been raised in 18 villages in partnership with fishing communities. Further, a coastal farming system involving mangrove plantations and aquaculture is becoming popular. Mangroves are also very efficient in carbon sequestration, thereby contributing to the maintenance of carbon balance in the atmosphere.

The other post-tsunami initiatives of MSSRF scientists include the establishment of coastal bio villages, which can enlarge opportunities for sustainable livelihoods. For example, tsunami-affected fisher-women were trained in a few villages in poultry farming, which has proved to be a highly remunerative occupation. Another programme involved the provision to the affected families of solar lamps to replace the smoky glow of kerosene lamps while going out in the sea.

During discussions with fisher families, a strong desire for opportunities for training in the science and art of sustainable fisheries was expressed. In response to this request, a Fish for All Research and Training Centre has been set up at Poompuhar with support from Tata Trusts. This unique field research and capacity-building centre will impart training, based on the pedagogic methodology of learning by doing, to fisherwomen and men in a holistic manner, ranging from fish capture or culture to fish processing and marketing. Training in all aspects of sustainable fisheries covering conservation, capture, consumption, and commerce will be imparted. The Poompuhar Fish for All Research and Training Centre is designed to foster a technological and management revolution in small-scale fisheries.

Another MSSRF initiative has been the establishment of computer-aided and internet-connected Village Resource Centres (VRCs) and Village Knowledge Centres (VKCs). The VRCs established with the help of the Indian Space Research Organisation have satellite connectivity and teleconferencing facilities. VKCs and VRCs, managed by trained local women and men, provide demand-driven and dynamic information. Synergy between the internet and mobile phones helps fishermen in catamarans to get the latest information on wave heights at different distances from the shoreline and on the location of fish shoals. This helps to allay fears and save time in harvesting fish. Recent developments in Information and Communication Technologies have opened up uncommon opportunities for helping small-scale fishermen to practise safe and sustainable marine fisheries. This is one of the fascinating and meaningful applications of mobile phone technology. Thus, coastal bio-shields, bio villages, and knowledge centres have become important tools for integrating ecological and livelihood security in a symbiotic manner in coastal areas. The proposal is, in cooperation with panchayati raj institutions, to train one woman and one man in every block as Climate Risk Managers; they should be well versed in disaster prevention, mitigation, and management. Coastal farm families were also affected by seawater intrusion as a result of the tsunami. For them, an agronomic rehabilitation package was introduced immediately.

As far back as 1991, MSSRF initiated an anticipatory research programme to meet the challenge of sea level rise, which involves the transfer of genes for seawater tolerance from mangroves to rice, pulses, and other coastal zone crops. This strategic research programme has led to the breeding of salt tolerant varieties of rice, which are undergoing tests as per prescribed regulatory procedures.

Impressive progress has been made by the State governments and non-governmental organisations in providing well-designed and hygienic homes to the affected families. The calamitous tsunami thus marked the beginning of a new chapter in the lives of those who lost their homes, fish boats, and much else. It is possible that global climate change will increase the frequency of such trials. Anticipatory action plans for managing the consequences of seawater intrusion in our coastal areas have become an imperative. At the same time, seawater is a valuable resource for raising salt-tolerant trees and crop varieties and fish in suitable agro-forestry and silvi-aquaculture systems. The 2004 tsunami has thus opened a new chapter in the lives of those who depend on the ocean for their livelihood as well as those who live near the sea and derive their income from a variety of opportunities, including farming, industry, and tourism.

Chapter 8

Now for the Evergreen Revolution

In India, farming is part of our culture. Seventy per cent of our population are engaged in farming. Half the world's farmers live in India or China: Every fourth farmer is Indian.

Famines were recurrent in India before Independence. Between 1870 and 1900, according to British records, 30 million people died of hunger and starvation. Nearly three million people died in the great Bengal famine — in what is now Bangladesh and India — in 1943–1944, just before Independence. After Independence, both Nehru and Indira Gandhi laid great emphasis on bringing more land under irrigation, in order to insulate our farming from being "a gamble on the monsoon", as Sir Albert Howard wrote in 1916. As every farmer knows, without water you can do nothing.

From the time I joined the Agricultural College at Coimbatore in Tamil Nadu in 1944, I have seen India's agricultural destiny transformed from being purely a gamble on the rain to being a gamble on the market. In 1950, our total food grain production was 50 million tonnes. In 2009–2010, it is estimated to be 218 million tonnes. Our average growth rate, particularly in the last 30 years, has been about 3 per cent per annum, which is above our population growth rate.

Twenty-five years ago we were number 25 in the world in wheat production, and now we are number two. We are number two in rice production and number one in milk production: All produced by very small farmers. To quote Mahatma Gandhi, "our production is production by the masses" — in contrast to the mass-production technology of the industrialised nations. What we need in India is not jobless growth but job-led economic growth: A human-centered kind of development. We need more farmers' farming and less factory farming.

How did this progress take place? The most important factor was the farmers' receptivity. Our farmers were thought of by Western writers as fatalistic, unlikely to respond to technology. They have proved to be like farmers anywhere else, with three determinants affecting their decisions — cost, risk and return.

The government had three major roles. One was technological: both National and State governments made large investments in agricultural research and education. We have a vast network of agricultural institutes and universities. And we have had the good fortune of close international partnership.

Technology alone is not adequate. Telling the farmer, "Grow this seed" has no particular meaning unless the seeds are available. Jalna, in Maharashtra, has become "the seed capital of India". Such services as the production of seed, irrigation, credit supply, fertilisers have been very important.

Government's third, and most important, contribution was a package of public policy, particularly in terms of agrarian reform and input–output pricing. Thirty years ago, the average Punjab farmer produced one tonne of rice per hectare, and kept 800 kg or so for his family. But if he can produce five tonnes, then he has four tonnes to sell and more cash in hand. The smaller the farm, the greater the need for a marketable surplus.

A small farm is ideal for intensive, precision agriculture. On the other hand, small farmers often cannot take risks, have no access to credit and are resource-poor. This is why public policy, such as the small farmers' programme, marginal farmers' programme and programmes for drought-prone areas, matters so much.

The milk revolution was partly achieved by technology, but mostly by institutional reform. The cooperatives gave the power of scale to the woman who had only three or four litres to sell per day.

Cooperatives, joint stock companies and other forms of organisation give small producers the power of scale both at the production end and at the post-harvest end. Modern ecological agriculture involves integrated pest management, integrated nutrient supply, scientific water management — "more crop per drop". None of this can be done by a single small farmer alone. It has to be done on an area basis.

Our population now exceeds one billion. Arable land is going out of agriculture all the time. Water resources are shrinking. Our groundwater is getting depleted and in many cases polluted. They are drawing water from the lower strata in West Bengal and Bangladesh with the result that arsenic is coming up. The greatest internal threat to agriculture is now the problem of conserving our soil fertility, land and water.

We are now in the phase which I call the evergreen revolution. This is where you have continuous advancement in productivity, but without associated ecological harm. It has three major preconditions. The first is a happy farming family. Scientists like me can give advice and materials, but the farmers are the ones who toil in the sun and rain and produce the food. So we should not only work for the consumer, but also think of the producer.

The second ingredient is a sustainable farming system, not based on one crop alone. Almost every farm in India has one or two cows or buffaloes, or small ruminants like goat or sheep. These are part of our life — crops, animals, fish, agro-forestry. If you go to Kerala every farm, every house, is a genetic garden: you will find jackfruit, pepper, coconut, up to 20 economic species in each garden. A sustainable farming system is the very foundation of organic agriculture — some degree of animal husbandry, composting, organic manure and also crop rotation. In the bio villages started by the JRD Tata Ecotechnology Centre of the MS Swaminathan Research Foundation (MSSRF), one crop is a very powerful nitrogen fixer called *Sesbania prostrata.* It fixes large quantities of nitrogen both in the stem and in the root. You may need to combine this with mineral fertilisers and chemical pesticides, in a way that maintains both environmental and social sustainability and economic viability.

The third precondition is sustainable food security. Although the Indian Government has more than 50 million tonnes of wheat and rice in its warehouses, over 250 million women, children and men still go to bed hungry. Therefore, jobs should be the bottom line of our agricultural policy. Where there is work, there is money, and where there is money there is food.

The JRD Tata Ecotechnology Centre tries to marry traditional wisdom, knowledge and technologies with the best in modern life. Knowledge is a continuum, every one of us leaves behind something new, and so will our children. You cannot freeze knowledge. So we have taken five different technologies: biotechnology, space technology (particularly in weather satellites), nuclear technology (particularly probes for underground mapping), information and digital technology (we have set up a series of information villages), and management technology. This last, in our definition, puts everything together into a management system for a farm which can be applied day by day.

Ecotechnology helps to bridge the divides in our country. We have found, for example, that bridging the digital divide in villages is a powerful method of bridging the gender divide. The people who are operating the knowledge centres in our information villages are women. Everybody comes to them for knowledge and this builds their self-esteem.

Everybody today talks about the globalisation of the economy. Everyone knows about anything which happens — particularly natural catastrophes — at the same time all over the earth. The global village in terms of information is a reality. But in economic terms it is a highly divided village.

We also know that our fates are intertwined ecologically. We can argue endlessly about who is responsible for mucking up the climate. But the fact remains that ecologically our fates are intertwined, and that is why we talk of our common future. But you cannot have a common future without a common present. In 1994 agriculture was introduced into the World Trade Agreement for the first time. It had five major components of importance to farmers: Access to markets, domestic support from governments, export subsidies, sanitary measures — using products which are completely free of *salmonella* and toxins — and Trade-Related Intellectual Property Rights (TRIPS) — patenting and protection of systems and plant varieties.

The Indian experience of this has been negative. We have no additional market access; in fact the market has even reduced in the Organisation for Economic Co-operation and Development (OECD) countries in the last few years. The domestic support we are given is a fraction of what is being given in the OECD countries. We do not give export subsidies.

Our sanitary measures are still poor, and this is bad for our own consumers. I tell my colleagues: "Don't only think that the foreign consumer doesn't want *Salmonella*, our people don't want it either". Quality is quality, and we should not think only of export-quality. I have been calling for all our agricultural universities to set up short non-degree courses for farmers in the Codex Alimentarius, put together by the Food and Agricultural Organisation (FAO) for food safety.

TRIPS has also worked against us so far. There are accusations of bio-piracy from "gene-rich" countries like ours towards the "technology-rich" countries. These are divisive forces. We hope that there will be a renegotiation of the World Trade Agreement on Agriculture.

Sarvodaya was a term coined by Gandhiji, meaning a win–win situation for all. I would like to see a sarvodaya world of farming emerge, where there is unity of purpose in spite of the diversity of methodologies, farming systems, climates, soil and needs.

Mahatma Gandhi said the pathway to achieve sarvodaya is *antyodaya* — attention to the poorest person. So if you want to have a sarvodaya farming world, then I think the more affluent members will have to pay attention to those who are economically, socially and ecologically handicapped. As Gandhiji said: "Before you do anything, ask yourself whether what you are about to do will help the poorest person you have seen in your life".

We have to marry ethics with economics and technology. The technological push must be matched by an ethical pull. If you do not have these two matching each other, then you cannot make sustainable progress.

Chapter 9

Evergreen Revolution and Sustainable Food Security

The Green Revolution, a term coined by William Gaud in 1968, is a process that leads to improved agricultural productivity.

I have used the term "evergreen revolution" to highlight the pathway of increasing production and productivity in a manner such that short- and long-term goals of food production are not mutually antagonistic, and to emphasise the need to improve productivity in perpetuity without associated ecological and/or social harm.

How do we achieve this evergreen revolution, i.e., a balance between human numbers and human capacity to produce food of adequate quantity, quality and variety? The growing damage to the ecological foundations essential for sustainable food security — land, water, biodiversity, forests and the atmosphere — is leading to stagnation in yields in Green Revolution areas. Climate change will compound such problems with adverse effects on temperature, precipitation, sea level and ultra-violet B radiation.

An analysis of food insecurity indicators in rural India carried out by the MS Swaminathan Research Foundation, with support from the World Food Programme (WFP), indicates that the Punjab–Haryana region — India's food basket — may become food-insecure in another 20 years. Indicators used in measuring sustainability of food security are: land degradation and salinisation, extent of forest cover, groundwater depletion and the nature of crop rotation. In all of these parameters, Punjab and Haryana occupy low positions. The common rice–wheat rotation has led to displacement of grain and fodder legumes capable of improving soil fertility. The current trend is towards non-sustainable farming resulting from land and water mining.

Forewarned is forearmed. What can we do to launch global agriculture on the pathway to an evergreen revolution, where advances in crop and farm animal productivity are not accompanied by either ecological or social harm?

Food security has three major dimensions:

- Availability of food — a function of production.
- Access to food — a function of purchasing power/access to sustainable livelihoods.
- Absorption of food in the body — determined by access to safe drinking water and non-food factors such as environmental hygiene, primary health care and primary education.

Capacity to support even the existing human and animal populations has been exceeded in many parts of the developing world. Hence, the future of food security depends upon population stabilisation, the conservation and care of arable land through attention to soil health and replenishment of fertility, and the conservation and careful management of all water sources so that more crop can be produced per drop of water.

Much of the degraded and desertified land belongs either to resource-poor families or constitutes over-used and over-grazed common property. Ownership patterns of land and water determine the feasibility of introducing integrated and sustainable land- and water-management systems. Even where land is individually owned, locally acceptable systems of social management may have to be introduced through legislation, education and social mobilisation. Women's access to land is also important. Water, particularly groundwater, should be a social resource and not private property. Creating an economic stake in conservation is vital for ensuring the sustainable use of natural resources.

Degradation and erosion of arable land and the depletion and pollution of water resources result in the loss of rural livelihoods.

This triggers unplanned migration of the rural poor to towns and cities, with proliferation of urban slums. The rise in the numbers of such environmental refugees threatens peace and security. There should be a monitoring mechanism for avoiding loss of rural livelihoods. Development programmes should strengthen linkages between ecological and livelihood security.

There are now unique opportunities for launching a food-for-sustainable-development initiative, in the form of a "grain for green" movement. Such a program could accord priority to:

1. Restoration of hydrological and biodiversity "hot spots", particularly in mountain ecosystems.
2. Coastal agro-aqua farms (planting of salicornia, mangroves, casuarina, palms, etc., along with coastal agriculture and aquaculture).
3. Water harvesting, watershed development, wasteland reclamation, and antidesertification measures.
4. Recycling of solid and liquid wastes and composting.
5. Agro-forestry and other sustainable land-use systems in the fields of resource-poor farmers.

A Global Food for Sustainable Development and Hunger Elimination Initiative could be launched by the International Alliance against Hunger, proposed by the Food and Agricultural Organisation (FAO). About 25 million tonnes of grains would provide nearly 100 million person-years of work designed to eliminate poverty-induced endemic hunger and at the same time restore and enhance environmental capital stocks.

Such food-for-ecodevelopment initiatives could be managed at the local level by Community Food Banks (CFBs) operated by women's self-help groups. Such CFBs can be designed to address concurrent issues relating to chronic, hidden and transient hunger. The merit of CFBs will be low transaction costs and transparency. They can also help to widen the food-security basket, thereby saving what could become "lost" crops. Where animal husbandry, including poultry farming, is important to provide additional income and nutrition to families living in poverty, CFBs could also operate feed and fodder banks.

It is the fundamental duty of the state as well as of the well-to-do sections of the population to confer on those who go to bed undernourished the right to food and thereby to opportunities to lead productive and healthy lives. Thanks both to the spread of democratic systems of governance at the grass-roots level and to technological advances, we now have a unique opportunity to foster a community-centered and controlled-nutrition security system. Such decentralised community management will help to improve delivery of entitlements, reduce transaction and transport costs, eliminate corruption and cater to the twin needs of introducing a lifecycle approach to nutrition security and meeting the challenge of seasonal fluctuation in nutritional status. If such CFBs are operated by women, this will help to bridge the gender divide in the area of nutrition.

Organic farming is gaining popularity among both farmers and consumers. Research should be intensified in several aspects to ensure that organic farming leads to higher productivity per unit of land and water used.

The earlier methods of soil-fertility management, like shifting cultivation, are no longer relevant today due to population pressure on land. Cereal-legume rotations and intercropping are important for replenishing soil fertility. Efficient green-manure plants like the stem-nodulating *Sesbania rostrata* and bio-fertilisers comprising efficient microorganisms have to be packaged in an integrated nutrient-supply system, which includes the application of compost, organic manures and plant residues. Inputs are needed to ensure outputs. For example, a tonne of rice needs at least 20 kg of nitrogen along with appropriate quantities of phosphorus, potassium and micronutrients. Research on soil-health management, in order to ensure adequate soil fertility for high productivity, should receive high priority.

All organic farmers should be provided with soil health cards to regularly monitor the physics, chemistry, microbiology and erodability of their soils. Care of soil health is fundamental to productive agriculture.

Sustainable organic farming will also need bioremediation agents that can help to improve soil health through the sequestration of salt, heavy metals and other yield-reducing constraints. A consortium of microorganisms, each capable of performing an important function like nitrogen fixation, phosphorus solubilisation, and/or sequestration of salts and pollutants, will be needed for each major agro-climatic and agro-ecological farming system.

The other area of research that is essential for sustained high productivity is integrated pest management involving concurrent attention to pests, diseases and weeds. For this purpose, there is need for a biosecurity compact that will help to manage not only pests, diseases and weeds, but also invasive alien species and mycotoxins in food. Sanitary and phytosanitary measures and Codex Alimentarius standards of food safety need to be integrated in organic production protocols.

As population pressure on land and water increases, there will be need for productive genotypes of crop plants that can perform well under conditions of soil salinity, alkalinity and acidity. Special genetic gardens will have to be established for halophytes and drought-tolerant genotypes. Also, suitable donors for tolerance of salinity and drought will have to be used in anticipatory breeding for adaptation to climate change and sea-level rise. Scientists at MSSRF have developed sea-water tolerant genotypes of rice, mustard and legumes using the mangrove species *Avicennia marina* as donor. Similarly, *Prosopis juliflora* is being used as a donor of genes for drought tolerance. Such pre-breeding work needs to be integrated with participatory breeding with farm women and men so that location-specific varieties can be developed. Genetic diversity is essential to avoid vulnerability to pests and diseases. Therefore, gene-deployment strategies will have to be developed jointly by scientists and farm families for each agro-ecological region. Successful organic agriculture will need a paradigm shift from purely experiment-station-based research to participatory research in farmers' fields.

Recent research at MSSRF has led to the isolation of a bacterial strain capable of fixing nitrogen and solubilising phosphate. *Swaminathania salitolerans gen. nov., sp. nov.* has been isolated from the rhizosphere, roots and stems of salt-tolerant wild rice associated with mangrove species. Field trials in rice using this microorganism are now in progress.

Sustainable organic agriculture will need more science, not less. Artificial barriers should not be created between scientific methods. It is important to harness all the tools that traditional wisdom and contemporary science can offer in order to usher in an era of biohappiness. The first requirement for biohappiness is nutrition and water security for all and forever. This is the challenge before all involved in organic farming and the seed industry.

The seed industry has a particularly vital role to play in ensuring genetic diversity in crop plants and in providing organic farmers with genotypes based on a pyramiding of genes for tolerance to major biotic and abiotic stresses. There is also need for greater attention to under-utilised or orphan crops, many of which are not only nutritious but also capable of performing well under fragile and rainfed environments.

In order to change the mindset relating to nutritious millets, the FAO should change the terminology from "coarse cereals" to "nutritious cereals". There is a need to reverse the narrowing of global food crops by including a wider range of cereals, millets, grain, legumes, vegetables and tubers in the diet. In the past, human communities depended upon several hundred species of plants for their nutrition and health security. Diversified farming systems and good dietary habits are essential to confer benefits both to the producer and to the consumer of organic farming products.

Production agriculture and forestry are the major solar energy-harvesting enterprises of the world. An evergreen revolution will help to optimise the production of farm commodities through a symbiotic interaction between solar and cultural energy. This is the pathway to sustainable food security and biohappiness.

Chapter 10

Priorities in Agricultural Research and Education

It was in 1968 that the term Green Revolution was coined by Dr. William Gaud of USA to describe advances in agriculture arising from productivity improvement. Even in 1968, I concluded that if farm ecology and economics go wrong, nothing else will go right in agriculture. I expressed my views in the following words in my lecture at the Indian Science Congress Session held at Varanasi in January 1968:

> Exploitive agriculture offers great dangers if carried out with only an immediate profit or production motive. The emerging exploitive farming community in India should become aware of this. Intensive cultivation of land without conservation of soil fertility and soil structure would lead, ultimately, to the springing up of deserts. Irrigation without arrangements for drainage would result in soils getting alkaline or saline. Indiscriminate use of pesticides, fungicides and herbicides could cause adverse changes in biological balance as well as lead to an increase in the incidence of cancer and other diseases, through the toxic residues present in the grains or other edible parts. Unscientific tapping of underground water will lead to the rapid exhaustion of this wonderful capital resource left to us through ages of natural farming. The rapid replacement of numerous locally-adapted varieties with one or two high-yielding strains in large contiguous areas would result in the spread of serious diseases capable of wiping out entire crops, as happened prior to the Irish potato famine of 1854 and the Bengal rice famine in

> 1942. Therefore, the initiation of exploitive agriculture without a proper understanding of the various consequences of every one of the changes introduced into traditional agriculture, and without first building up a proper scientific and training base to sustain it, may only lead us, in the long run, into an era of agricultural disaster rather than one of agricultural prosperity.

The above analysis led me to coin the term "evergreen revolution" to describe the enhancement of productivity in perpetuity without associated ecological harm. The pathways to an evergreen revolution are either organic farming or green agriculture. Green agriculture involves the adoption of environment friendly practices like integrated natural resources management and integrated pest management. It is our sacred duty to conserve and enhance the ecological foundations such as soil, water and biodiversity essential for sustained advances in agricultural productivity and profitability.

The present decade may mark the beginning of a new climate era, characterised by extreme and often unpredictable weather conditions and rise in sea levels. The recent Climate Conference in Copenhagen unfortunately failed to get a global commitment to halt economic growth based on high carbon intensity. The Climate Conference due to be held in Mexico in December 2010 will probably generate the political commitment essential to restrict the rise in global mean temperature to not more than 2°C, as compared to the mean temperature of today. Even a 2°C rise will adversely affect crop yields in South Asia and Sub-Saharan Africa, which already have a high degree of prevalence of endemic hunger. It will also lead to the possibility of small islands getting submerged. The greatest casualty of climate change will be food, water and livelihood security. Farmers of the world can help to avoid serious famines by developing and adopting climate-resilient farming systems. 2010 has been declared by the United Nations as the International Year of Biodiversity. As I have said earlier, biodiversity is the feedstock for a climate-resilient agriculture. We should therefore redouble our efforts to prevent genetic erosion and to promote the conservation and sustainable and equitable use of biodiversity.

2010 will also witness a major conference at the United Nations Headquarters in New York to review the progress made since the year 2000 in achieving the UN Millennium Development Goals. The first among these goals is reducing hunger and poverty by half by 2015. Unfortunately the number of hungry children, women and men, which was 800 million in 2000, is now over a billion. This is partly due to a rise in food prices, thereby making it difficult for the poor to have access to balanced diets at affordable prices.

Adaptation to Climate Change

A group of scientists led by the MS Swaminathan Research Foundation (MSSRF) have undertaken studies during the last five years in Rajasthan and Andhra Pradesh on climate change adaptation measures. The districts chosen were Udaipur in Rajasthan and Mehabubnagar in Andhra Pradesh. The approach adopted was to bring about a blend of traditional wisdom and modern science. The participatory research and knowledge management systems adopted under this programme during the past five years have provided many useful insights for developing a climate-resilient farming and livelihood security system. Five of the meaningful adaptation interventions have been the following:

- **Water conservation and sustainable and equitable use:** Families in the desert regions of Rajasthan have long experience in harvesting every drop of rainwater and using it economically and efficiently both for domestic and agricultural use. The traditional methods were reinforced with modern scientific knowledge, like the gravity flow method of water management.
- **Promoting fodder security:** Livestock and livelihoods are intimately related in arid and semi-arid areas. The ownership of livestock is also more egalitarian. The sustainable management of common property resources, particularly pasture land, is essential for ensuring fodder security. Therefore, high priority was given to the regeneration of pasture land and the equitable use of grazing land.

- **More crop and income per drop of water:** In areas where water for irrigation is the constraint, it is important that agronomic techniques which can help to increase yield and income per drop of water are standardised and popularised. One such method introduced under this project is the System of Rice Intensification (SRI). SRI was popularised in Andhra Pradesh, since this system of water and crop management helps to reduce irrigation water requirement by 30 to 40 per cent. This method thus helps to avoid the unsustainable exploitation of the aquifer.
- **Weather information for all and climate literacy:** What farmers need is location-specific meteorological information at the right time and place. Generic weather data will have to be converted into location-specific meteorological advice. For this purpose, mini-agro-meteorological stations managed by the local community were established. This has helped to impart climate literacy as related to food, water and livelihood security.
- **Strengthening community institutions:** Effective implementation of adaptation measures will need active group cooperation and community participation. Steps were taken to involve the grass–root democratic institutions like panchayats and gram sabhas. Also, Smart Farmers' Clubs were organised to give the power of scale in water harvesting, soil health management and other adaptation measures undertaken by farmers with smallholdings.

These interventions were supported by training, skill development, education and social mobilisation. A Training Manual was prepared by MSSRF for training one woman and one male member of every panchayat in the art and science of managing weather abnormalities, making them local-level Climate Rick Managers.

The work has highlighted the need for location-specific adaptation measures and for participatory research and knowledge management. The adaptation interventions have also highlighted the need for mainstreaming gender considerations in all interventions. Women will suffer more from climate change, since they have been traditionally in charge of collecting water, fodder and fuel wood, and have been shouldering the responsibility for the care of farm animals as well as for post-harvest technology. All interventions should therefore be pro-nature, pro-poor and pro-women.

The last five years have been an extremely rewarding learning period. The results and experience have shed light on the way forward. It is clear that to promote location-specific and farmer-centric adaptation measures, India will need a Climate Risk Management Research and Extension Centre at each of the 127 agro-ecological regions in the country. Such centres should prepare Drought, Flood and Good Weather Codes that can help to minimise the adverse impact of abnormal weather and to maximise the benefits of favourable monsoons and temperature. Risk surveillance and early warning should be the other responsibilities of such centres.

The work done so far has laid the foundation for a climate-resilient agriculture movement in India. The importance of such a movement will be obvious considering the fact that 60 per cent of India's population of 1.1 billion depend upon agriculture for their livelihood. In addition, India has to produce food, feed and fodder for over 1.1 billion human and over a billion farm animal populations.

Challenges Ahead

2010–2011 is a watershed year in the history of Indian agriculture. Producing food in adequate quantities and making them available at affordable prices will be the greatest challenge during this year. Also, our food security should be built on the foundation of home-grown food, since agriculture is the backbone of the livelihood security system of nearly 700 million people in the country. Nearly 60 per cent of the cultivated area is rain-fed and these are the areas where pulses, oilseeds and other crops of importance to nutrition security, such as millets, are grown. I need hardly emphasise that India is the home for the largest number of malnourished children, women and men in the world. The majority of the malnourished are producer-consumers (i.e., farmer-consumers) and landless labour. Increasing the productivity and profitability of small farms is the most effective method of achieving the UN Millennium Development Goal No. 1 — reducing hunger and poverty.

Road Map

A road map for our agricultural renaissance and agrarian prosperity was presented by the National Commission on Farmers (NCF) in five reports presented between 2004 and 2006. The reports are yet to be printed, let alone implemented. For example, 70 per cent of India's population do not find a place in the Padma awards announced on 26 January each year, although the NCF had stressed the need for according social prestige and recognition to farmers through such gestures. Farming, particularly in the heartland of the Green Revolution comprising Punjab, Haryana and Western Uttar Pradesh, is in deep ecological and economic crises. No wonder over 40 per cent of the farmers surveyed by the National Sample Survey Organisation wish to quit farming, if there is another option.

Some of the areas needing immediate attention and action are discussed below.

Defending the gains already made in the Green Resolution areas through conservation farming, involving concurrent attention to soil health enhancement, water conservation and effective use, biodiversity protection and launching of a climate-resilient agriculture movement is an urgent task. These are the areas which feed the public distribution system. NCF had recommended the allocation of Rs. 1000 crore for this purpose. Expenditure in this area will also come under the Green Box provision of WTO. Climate-resilient agriculture will involve shifting attention to per day rather than per crop productivity.

There is immense untapped production potential in eastern India, the sleeping giant of Indian agriculture — Bihar, Chattisgarh, Jharkhand, eastern Uttar Pradesh, West Bengal, Assam and Odisha. A large number of government schemes with substantial financial outlays, like the Rashtriya Krishi Vikas Yojana, the Food Security Mission, and the National Horticulture Mission exist, but are not making the desired impact on the productivity and production of small farmers. A well-planned movement to bridge the yield gap needs to be initiated with the active involvement of farming families and gram sabhas.

The gap between potential and actual yields with the technologies on the shelf ranges from 200 to 300 per cent in these areas. Prime Minister Rajiv Gandhi initiated a dry land farming revolution in these areas through the Pulses and Oilseeds Missions, but the end-to-end approach he had designed was soon given up and there has been a reversion to the business-as-usual approach. I suggest that during 2010–2011, 60,000 pulses and oilseed villages may be organised in rain-fed areas, to mark the 60th anniversary of our Republic. In each of these villages, there should be a lab to land programme organised by the Indian Council of Agricultural Research and State agricultural universities. Such villages can be developed with the help of gram sabhas and with the active involvement of farm scientists with the requisite knowledge and experience. The Mahatma Gandhi National Rural Employment Guarantee Act (MGNREGA) workers can help in water harvesting, watershed management and soil health enhancement. Integrated attention to conservation, cultivation, consumption and commerce should be paid. Assured and remunerative marketing will hold the key to stimulating and sustaining farmers' interest. Today, the consumer is paying very high prices for pulses, but the producer lives in poverty.

Conferring the economy and power of scale to farm families with smallholdings is the most serious challenge facing our agriculture. Farm size is declining and 70 per cent of farmers cultivated less than 1 ha in 2003, compared with 56 per cent in 1982. Cooperative farming has been successful in the dairy sector in Gujarat and a few other States. It has not been successful in crop husbandry, although Andhra Pradesh has recently initiated a programme for promoting farm cooperatives. There is increasing feminisation of agriculture with 83 per cent of rural female workers engaged in work related to crop and animal husbandry, fisheries and forestry. Gender specific needs of women farmers, including credit, technology, training and support services like crèches and daycare centres, are urgently needed. It is a matter of satisfaction that the *Mahila Kisan Sashakthikaran Pariyojana* started by MSSRF in Vidarbha three years ago for the skill and management empowerment of women farmers, including the widows of farmers who had committed suicide, was elevated into a national programme in the Union Budget for 2010–2011, with an initial allocation of Rs. 100 crore.

Nearly 70 per cent of our population is below the age of 35 and 70 per cent of them live in villages. The future of our agriculture will depend upon attracting and retaining youth in farming. This is one of the principal goals of the National Policy for Farmers (2007). There are several government projects, which if revamped and revitalised, can help to make farming as a profession attractive to educated youth. A new programme for youth in agriculture may be initiated by integrating several ongoing schemes like the Small Farmers' Agri-Business Consortium (SFAC), Agri-Clinics, Agri-Business Centres, Food Parks, etc. This will help to stimulate the formation of Young Farmers' Self-Help Groups. SFAC could be developed into a Young Farmers' Agri-Business Consortium, bringing together all relevant programmes.

Food and water security will be the major casualties of a rise in mean temperature, monsoon uncertainty, drought, floods and sea level rise. Some of the steps which could be initiated during 2010–2011 are:

- Promote a water conservation pond as well as a biogas plant in every farm, wherever there are farm animals.
- Plant one billion fertiliser trees which can serve as soil carbon banks, enrich soil fertility and enhance farm productivity. Funds for this purpose (Rs. 13,000 crore) are available with the Ministry of Environment and Forest under the Compensatory Afforestation Fund Management and Planning Authority (CAMPA).
- Establish farmer participatory research and training centres for climate change management in each of the 127 agro-climate zones of the country. Such centres will train at least 1 woman and 1 man in every panchayat as Climate Risk Managers. The centres can be located in either Agricultural and Animal Sciences universities or Krishi Vigyan Kendras or ICAR institutes.
- Build mangrove and non-mangrove bioshields along the coast. These are essential for reducing damage from sea level rise,

cyclones and tsunamis. Along with the bioshields, 1000 seawater farming demonstrations can be organised. Seawater is a social resource, as stressed by Mahatma Gandhi when he launched the Salt Satyagraha. Seawater farming will involve the establishment of agri-aqua farms. The farmer participatory demonstrations could be organised along the Indian coast as well as in the Lakshadweep and Andaman group of islands.

During this year, we should begin establishing ultra-modern grain storages at least in 50 locations in the country, each with a storage capacity of a million tonnes of foodgrains (i.e., a 50 million tonne storage grid). The government should remain at the commanding height of the food security system.

2010 is a do or die year for Indian agriculture. If we do not take steps to address the serious ecological, economic and social crises facing our farm families, we will be forced to support foreign farmers, through extensive food imports. This will result in a rise in food inflation, increase the rural–urban and rich–poor divides and allow the era of farmers' suicides to persist. On the other hand, we have a unique opportunity for ensuring food for all by mobilising the power of youth and women farmers and by harnessing the vast untapped yield reservoir existing in most farming systems through synergy between technology and public policy.

Overcoming hidden hunger caused by micronutrient deficiencies like iron, iodine, zinc, vitamin A and vitamin B12 can be achieved by growing and consuming appropriate local vegetables and fruits. There is a horticultural remedy for every nutritional malady. *Moringa*, which is a jewel in the horticultural crown, is an example.

Urban and non-farming members of the human family should realise that we live on this planet as the guests of sunlight and green plants, and of the farm women and men who toil in the sun and rain, and day and night, to produce food for over 6 billion people, by bringing about synergy between green plants and sunlight. Let us salute the farmers of the world and help them in achieving the goal of a hunger-free world, the first among the UN Millennium Development Goals.

This decade will show that the future belongs to nations with grains and not guns. Human destiny during this decade and beyond will be shaped by farm women and men. This decade will thus be the Decade of Farmers.

Chapter 11

Achieving Food Security in Times of Crisis

Food security involves physical, economic, social and environmental access to a balanced diet and clean drinking water to every child, woman and man. Physical access is a function of the availability of food in the market and is related to both in-country production and imports, when needed. Economic access is related to purchasing power and employment opportunities. Social access is conditioned by gender equity and justice. Environmental access is determined by sanitation, hygiene, primary healthcare and clean drinking water. Thus, both food and non-food factors determine food security.

The world is facing a hunger crisis. In spite of the highest priority accorded to hunger elimination among the UN Millennium Development Goals (UN-MDGs), the Food and Agricultural Organisation (FAO) estimates that the number of people going to bed hungry is increasing. When UN-MDGs were adopted in 2000, about 820 million were estimated to be under-nourished. Now, it is over a billion. Why are we in this condition?

Aristotle said long ago that the soil is the stomach of the plant. Exploitative agricultural practices lead to soil mining and damage to the physical, chemical and microbiological properties of the soil. Every farm family should have a soil health card giving integrated information on all aspects of soil health, like organic matter status, macro- and micro-nutrient availability and the hydraulic conductivity of the soil. A national land care movement should deal with both the conservation of prime farm land for agricultural purposes and the prevention of soil erosion and degradation. The fertility of waste or wasted land should be restored.

Building a sustainable water security system involves concurrent attention to supply augmentation and demand management. Supply augmentation involves harnessing all the major sources of irrigation water — rain, ground, surface, effluents and waste water and seawater. Rainwater harvesting through a pond in every farm must become a way of life. Seawater constitutes over 97 per cent of the water resources available in our planet. There is vast scope for seawater farming through agri-aqua farms. Conjunctive use of water like fresh water and treated industrial effluents should become institutionalised. Industry should give back the water it consumes in a good condition. Demand management in agriculture should come from the adoption of "more crop and income per drop of water" techniques. Agronomists should indicate in their publications not only yield per hectare, but also yield per unit of water. Micro-irrigation methods need to become universal.

Biodiversity loss and damage to ecosystem services is taking place at an alarming rate. This has serious implications in relation to our capacity to deal with the new challenges arising from climate change and transboundary pests. The loss of every gene and species limits our options for the future, particularly when recombinant Deoxyribonucleic Acid (DNA) technology affords opportunities to create novel genetic combinations capable of conferring resistance to abiotic and biotic stresses.

An institutional method to address environmental threats to food security is the organisation of community-managed food and water security systems at the village level. This will comprise Field Gene Banks through *in situ* on-farm conservation of local landraces, Seed Banks for ensuring the availability of seeds during times of drought and flood, Grain Banks involving storage of local food crops (often belonging to the category of orphan crops) and Water Banks in the form of ponds and reservoirs capturing rain water. Thus, conservation, cultivation, consumption, and commerce can be linked into a food security continuum. A reason why malnutrition is increasing in the world is the centralised approach to both analysis and action. A decentralised, community-centered approach to food security will help us to reach our nutrition goals speedily and surely.

The cost–risk–return structure of farming determines the decisions of farmers with reference to the choice of crops and investment on inputs. Input costs are going up partly due to the escalation in the price of petroleum products. Output prices are not increasing in tandem with a rise in the cost of production. Due to inadequate availability of institutional credit and effective insurance, small farmers get caught in a debt trap, with much of the borrowing coming from private money lenders at very high interest rates. On one hand, public policies in the field of agriculture should give over-riding priority to safeguarding and improving the ecological foundations essential for sustainable agriculture, and on the other, assured and remunerative marketing opportunities.

The social, economic, environmental and gender dimensions of equity must receive integrated attention. An area in intra-generational equity which needs urgent attention is the elimination of maternal and foetal undernutrition resulting in the birth of children with Low Birth Weight (LBW). Such LBW children suffer from several handicaps including impaired cognitive abilities. At the other end is the growing damage to our life support systems of land, water, biodiversity and climate, leading to reduced opportunities for a healthy and productive life to the children yet to be born.

A method of overcoming problems in the areas of environmental, social and gender inequity is to subject all the development programmes to a matrix analysis designed to ascertain whether the programme is pro-nature, pro-poor and pro-woman. Such an orientation is also essential in the area of technology development and dissemination.

The famine of jobs or purchasing power is often the cause of famine of food at the household level. Modern industry often leads to jobless economic growth. Agriculture — including crop and animal husbandry, fisheries, forestry, agro-forestry, agro-processing and agri-business — promotes job-led growth. Crop-livestock integrated farming systems enhance both income and nutrition security. In the developing countries of the Asia-Pacific region, what we need is job-led economic growth, so that the goal of food for all coupled with human dignity can be achieved. The economics of human dignity demands that everyone should have an opportunity to earn his/her daily bread.

There are several successful models of promoting job-led economic growth in this region. One model relates to the successful experience in China of promoting higher small farm productivity and profitability, on one hand, and opportunities for skilled and remunerative non-farm employment through Township–Village–Enterprises (TVE), on the other. This two-pronged strategy has helped China achieve both high farm productivity and impressive manufacturing capacity. "Jobs for All" then becomes a reality.

The other model, developed at the MS Swaminathan Research Organisation (MSSRF), is known as the "Bio village" model of human-centered development. The Bio village model involves the following concurrent steps:

- Conservation and enhancement of the ecological foundation for sustainable agriculture, with particular attention to soil health care, rainwater harvesting and efficient water use, conservation as well as sustainable and equitable use of biodiversity, climate risk management, and the protection and development of village common property resources.
- Improvement of on-farm productivity based on evergreen revolution principles, which help to enhance farm productivity in perpetuity without associated ecological harm, through mainstreaming ecological principles in technology development and dissemination.
- Generation of skilled and market-driven non-farm employment opportunities through improved post-harvest technology and value addition to primary products.
- Greater investment of technology and finance in the processing, storage and marketing of foodgrains.

The Biovillage Council, which manages the biovillage activities through group cooperation, ensures that every adult in the village has an opportunity for a healthy and productive life. Each Biovillage Council develops a strategy for energy security involving a feasible and affordable blend of renewable and non-renewable sources of energy. Among the renewable sources solar, wind, biogas and biomass are particularly important.

Starting from the industrial revolution in Europe nearly four centuries ago, technology has been a major factor in North–South, rich–poor, rural–urban and gender divides. If technology has been the primary cause of such divides, we should now enlist technology as an aid to bridging the divides. An important requirement for promoting the "Bridging the Divides Movement" is knowledge and skill empowerment. Harnessing modern Information Communication Technologies (ICT) is a powerful method of empowerment of rural communities. The Village Knowledge Centre movement, launched in India by MSSRF in partnership with a multi-stakeholder National Alliance for Village Knowledge Revolution, is based on the principles of community ownership, demand driven and dynamic information, use of local language and capacity building. Capacity building and content creation are two key elements of this programme.

Biotechnology is becoming an important tool in creating novel genetic combinations. Action is needed at two ends of the spectrum for harnessing novel genetic combinations to meet current and future challenges arising from global warming and climate change. First, DNA Clubs should be organised in village schools to spread genetic literacy. Second, each nation should have a statutory, professionally-led National Biotechnology Authority, emphasising "the economic well-being of farm families, food security of the nation, health security of the consumer, protection of the environment, biosecurity of the country and the security of national and international trade in farm commodities" (Report of the M.S. Swaminathan Committee 2004). Developing countries should institute regulatory procedures to ensure the safe and responsible use of biotechnology, particularly recombinant DNA technology. In India, a National Biotechnology Regulatory Authority is being created through an Act of Parliament.

The need for adopting the methods of an evergreen revolution by mainstreaming the principles of ecology in technology development and dissemination has become very urgent. There are two major pathways to fostering an evergreen revolution. The first is organic farming. Productive organic farming needs considerable research support, particularly in the areas of soil fertility replenishment and plant protection. Soils in most parts of India lack organic matter and are also deficient both in macro- and micro-nutrients. A majority of farmers cultivate one hectare or less. Crop-livestock integrated farming will help to build soil fertility but most small farm families have only 1 or 2 farm animals like cows, buffaloes and bullocks. Green manure crops and fertiliser trees can help to build soil fertility. Also, commercially viable organic farming methods will spread only if there is a premium price for organic products. Organic farming should be promoted in the case of vegetable and fruit crops and medicinal plants, where the danger of pesticide residues should be avoided.

The other pathway to an evergreen revolution is green agriculture. In this case, ecologically sound practices like conservation farming, integrated pest management, integrated nutrient supply and natural resources conservation and enhancement, are promoted. Green agriculture techniques could include the cultivation of crop varieties bred through the use of recombinant DNA technology, in case such varieties have advantages like resistance to biotic or abiotic stresses, or other attributes like better nutritive quality. In organic farming, the cultivation of genetically modified crops is prohibited. The cultivation of varieties bred with the help of molecular marker-assisted selection is however allowed.

For resource-poor farmers, green agriculture is the method of choice for producing more in an environmentally benign manner. The smaller the farm, the greater is the need for marketable surplus. Research on efficient micro-organisms, which can help to build soil fertility, as well as fertiliser trees like *Faidherbia albida* will help both organic farming and green agriculture. The National Commission on Farmers (NCF) (2006) recommended the initiation of a conservation farming movement in the heartland of the Green Revolution in order to halt the damage now occurring to the ecological foundations essential for sustainable agriculture.

Despite the large number of nutrition safety net programmes introduced by the Central and State Governments from time to time, India still remains the home for the largest number of malnourished children and adults in the world. We should ask why we are in this regrettable and unacceptable situation. The answer lies in the basic structure of our consumption pattern.

Nearly two-thirds of our population live in rural areas. A majority of them are small and marginal farmers and landless labour. They fall under the category of producer–consumer. We have thus two categories: About 700 million producer–consumers and about 400 million consumers. In industrial countries, consumers will be about 97 per cent and producer–consumers will be about 3 per cent. Therefore, widespread malnutrition and endemic hunger will persist in India unless the producer–consumer can consume a balanced diet. This situation also prevails in most countries in the Asia-Pacific region. This will call for higher small farm productivity and profitability, on an environmentally sustainable basis. An evergreen revolution accompanied by a small farm management revolution are hence vital components of a freedom from hunger movement. How can we develop a sustainable and equitable food security system?

As pointed out earlier, food security at the level of each individual child, woman and man involves physical, economic and social access to balanced diet, including the needed macro- and micro-nutrients, safe drinking water, primary healthcare, sanitation, and environmental hygiene. Thus, concurrent attention is needed to both food and non-food factors. Any national legislation relation to food security should deal with production, access and absorption in a holistic manner.

Indian agriculture is at the crossroads. Our population may reach 1750 million by 2050. Per capita crop land will then be 0.089 ha and per capita fresh water supply will be 1190 m^3/year. Foodgrain production must be doubled and the area under irrigation should go up from the current 60 million ha to 114 million ha by 2050. Degraded soils should be restored through increase in carbon pools in soils. How are we going to achieve a match between human numbers and human capacity to produce adequate food for all? To quote Edward O Wilson (2002) in *The Future of Life*:

> The problem before us is how to feed billions of new mouths over the next several decades and save the rest of life at the same time

> without being trapped in a Faustian bargain that threatens freedom from security. The benefits must come from an evergreen revolution (as proposed by M.S. Swaminathan). The aim of this new thrust is to lift production well above the levels attained by the Green Revolution of the 1960s, using technology and regulatory policy more advanced and even safer than now in existence.

With the spread of democratic systems of governance in most parts of the world, a world without hunger is an idea whose time has come. Access to a balanced diet and clean drinking water must be a fundamental right of every human being. This will call for a shift from a charity-based approach to hunger elimination to a rights-based one. The Government of India is currently developing legislation to ensure food security for all. Such a National Food Security Act, to be effective, should deal with food availability, access and absorption in an integrated manner. Food availability can be ensured by launching a "bridge the yield gap" movement, which is designed to help in narrowing or eliminating the gap between potential and actual yields through packages of technology, services and public policies.

Food access can be ensured through making food available at affordable cost and by generating sustainable livelihood opportunities in the farm and off-farm sectors. A rights-based approach to access can provide for common and differentiated entitlements. The common entitlement should aim to ensure adequate availability of food in the market coupled with an effective public distribution system from which will enable all citizens to access essential quantities of staple grains at a reasonable price. Differentiated entitlement will refer to providing food at low prices to the socially and economically underprivileged sections of society. Thus, there will be universal access to the needed calories and proteins, making the goal of food for all a reality.

The National Food Security Act, in addition to aiming to end poverty induced protein-energy malnutrition, should also provide for the following:

- Elimination of hidden hunger caused by the deficiency of micronutrients like iron, iodine, zinc, vitamin A and vitamin B12

through a food-cum-fortification approach. In particular, emphasis should be placed on providing horticultural remedies for the nutritional maladies prevailing in an area, based on local foods.
- Provision of clean drinking water to ensure food assimilation in the body.
- Attention to non-food factors like primary health care, environmental hygiene and sanitation.
- Launch of a nutrition literacy movement and training one woman and one man in every village as "Hunger Fighters".

In the ultimate analysis we will succeed in achieving food security in an era of global change, only through a well planned and concerted endeavour at the global, national and local levels. Centralised goals and resource allocation should be coupled with decentralised planning and action. Community food and water security systems involving the establishment of local-level gene, seed, grain and water banks will facilitate both an evergreen farm revolution and sustainable food and nutrition security. Such local level food security systems will also help to enlarge the shrinking food basket by including a wide range of millets, legumes and tubers in the diet.

FAO is the flagship of the global resolve to end hunger. The Asia-Pacific Region is the home of the largest number of undernourished children, women and men. Hunger can be overcome if there is the requisite fusion of professional skill, political will and action, farmers' enthusiasm and, above all, people's participation. The FAO Regional Office for the Asia Pacific Region has the unique opportunity for promoting a Food Security Symphony to generate the needed degree of convergence and synergy among the numerous nutrition safety net programmes in operation in our region.

Mahatma Gandhi said in 1948: "God is bread to the hungry". Let us resolve to work together to ensure that every home in our region is blessed by the God of Bread.

Chapter 12

Synergy between Food Security Act and Mahatma Gandhi National Rural Employment Guarantee Act

In its latest election manifesto, the Congress pledged to "enact a Right to Food Law that guarantees access to sufficient food for all people, particularly the most vulnerable sectors of society". It further pledged that "every family below the poverty line either in rural or urban areas will be entitled by law to 25 kg of rice or wheat per month at Rs. 3 per kg". Also promised were subsidised community kitchens in all cities for homeless people and migrants with Central government support. And, "along the lines of the Mahatma Gandhi National Rural Employment Guarantee Act (MGNREGA) we will enact a National Food Security Act".

Such an act will meet a goal set by Mahatma Gandhi for independent India: "the god of bread" should bless every home and hut. There is an unacceptable extent of under-nutrition and malnutrition in India, which occupies a shameful position in all indices relating to hunger. A large segment of the chronically undernourished belongs to families of small and marginal farmers and landless labour. The position is serious in the case of women and children. Because of maternal and foetal undernutrition and malnutrition, nearly every fourth child born is underweight. Such low birthweight children suffer many handicaps including impaired cognitive ability. Thus, poor children are denied even at birth an opportunity for the full expression of their innate genetic potential for mental and physical development. This is inexcusable in a democratic society.

The successful implementation of MGNREGA and the Right to Information Act indicates that the climate is conducive for a far-reaching, rights-based legislation to eliminate hunger and deprivation. Supreme Court rulings reinforce the view that the right to food is basic to achieving the right to life enshrined in Article 21 of the Constitution. One of the terms of reference the United Progressive Alliance (UPA) government set in 2004 for the National Commission on Farmers (NCF) was to "work out a comprehensive medium-term strategy for food and nutrition security in the country in order to move towards the goal of universal food security over time". NCF held consultations all over India on the pathways to a nutrition-secure India. Its report was submitted on 4 October 2006.

By definition, food security involves every individual gaining physical, economic, social and environmental access to a balanced diet that includes the necessary macro- and micronutrients, safe drinking water, sanitation, environmental hygiene, primary healthcare and education so as to lead a healthy and productive life. The food should originate from efficient and environmentally benign production technologies that conserve and enhance the natural resource base of crops, farm animals, forestry, inland and marine fisheries.

Such a holistic definition requires concurrent attention to the following aspects, too:

Food availability: This is a function of home production or, where absolutely essential, imports.

Food access: This is a function of livelihood opportunities and purchasing power. As early as in 1859, a Famine Commission appointed by the colonial government said: "Indian famines are not famines of food, but of works; where there is work there is money and where there is money there is food". This is why Mahatma Gandhi said in 1946 at Noakhali: "To a people famishing and idle, the only acceptable form in which God can dare appear is work and promise of food as wages".

Food absorption: The utilisation of food in the body will depend on non-food factors such as safe drinking water, environmental hygiene, primary healthcare and access to toilets. Therefore, while developing legislation for food security, food and non-food factors will have to be considered together on the following lines.

Food availability: The government has initiated programmes to increase food production, such as the Rashtriya Krishi Vikas Yojana, the Food Security Mission, and the National Horticulture Mission. Food availability should also relate to macro- and micronutrients. In addition to protein calorie undernutrition, there is severe micronutrient malnutrition, as for example, of iron, iodine, vitamin A, vitamin B12 and zinc, leading to hidden hunger. The National Horticulture Mission provides an opportunity to introduce horticultural remedies to nutritional maladies. All that is needed is mainstreaming the nutritional dimension in designing the horticulture programme in each agro-climatic area.

The other areas which will need attention are: widening the food basket by including local grain varieties like ragi, jowar and millets in the public distribution system; the promotion of community gene, seed, food and water banks in each village, and the establishment of community kitchens modelled on the Indira Gandhi Community Kitchen organised years ago in Pune. These are particularly effective in combating malnutrition in urban areas. The widening of the food basket by including millets, legumes and tubers, which have greater tolerance to adverse conditions, is important in the context of climate change.

Food access: The Congress manifesto has said 25 kg of rice or wheat would be provided each month to economically underprivileged families at Rs. 3 per kg. With the initiation of MGNREGA, the minimum purchasing power for food security is being created in families living below the poverty line. By adopting the support price policy recommended by NCF, that is, C2 (total cost of production) plus 50 per cent as has been done in the case of wheat, the purchasing power of small and marginal farmers can be improved. Universalisation of the Public Distribution System (PDS) is an idea whose time has come, since there are adequate grain stocks.

Food absorption: Here, the schemes dealing with drinking water, sanitation, environmental hygiene and so on should be brought together. The Total Sanitation Programme and the Rajiv Gandhi

Drinking Water Mission, if implemented with community participation through panchayats and nagarpalikas, will make a difference in promoting effective absorption of food in the body, particularly among children.

With such a holistic approach, chronic, hidden and transient hunger can be addressed in a cost-effective and meaningful manner. To provide political oversight and to foster a pan-political approach in matters relating to food security, NCF recommended the establishment of a National Food Security and Sovereignty Board under the chairmanship of the Prime Minister and with members drawn from different political parties, Union Ministers and Chief Ministers. Such a political support and oversight body should become an integral part of the legislation.

The MGNREGA, which came into force in February 2006, now covers all of rural India. It has generated over 450 crore person-days of employment, a major share going to women and Scheduled Caste and Scheduled Tribe families. Over Rs. 35,000 crore has been paid as wages. The priorities of the work to be undertaken include watershed management and water conservation, drought-proofing, flood protection, land development, minor irrigation and rural connectivity. Such work is important to strengthen the ecological foundations of sustainable agriculture. MGNREGA is probably the world's largest ecological security programme. A major weakness has been the absence of effective technical guidance and support from agricultural and rural universities and institutes. The Union Ministry of Rural Development has taken steps to achieve convergence of brain and brawn, by enlisting the support of Ministries and Departments. Such convergence of expertise for sustainable development will help to enhance farm productivity without causing ecological harm.

What is now needed is a similar convergence for human development at MGNREGA sites. India occupies the 132nd position among 179 countries in the United Nations Development Programme's 2009 Human Development Index. That position may worsen. MGNREGA workers represent some of the most economically and socially underprivileged sections. Mostly, these workers are undernourished, with poor opportunities for healthcare. Hence, there is a need to bring about a coming together of child care, nutrition, health (Rural Health Mission) and education programmes at MGNREGA sites. Education can be imparted in the evenings, using the joyful learning techniques available in computer-aided literacy programmes. Such a convergence in sustainable development along with convergence in human development will help to convert Gandhiji's concept of "bread with human dignity" into reality.

There is also a need to raise the self-esteem of MGNREGA workers, making them feel proud of the fact that they are engaged in checking eco-destruction. Due recognition could be given to the MGNREGA groups that have done outstanding work in water harvesting, watershed development and soil healthcare with "Environment Saviour Awards". This will help spread awareness of the critical role MGNREGA workers play. To begin with, there could be 10 awards covering distinct agro-climatic zones, each worth Rs. 10 lakh. Since these will be group awards, the money could be divided among the workers, depending on how long they have worked. MGNREGA will then help to improve both food security and the country's position in the human development index.

There are uncommon opportunities to erase India's image as the land of the poor, hungry and illiterate. To utilise them, an important requirement is a change in the mindset from patronage to partnership and from undervaluing the human resource to considering our youthful population as our greatest asset. Mahatma Gandhi once said that the divorce between intellect and labour is the bane of rural India. MGNREGA provides a unique opportunity to achieve a marriage between intellect and labour or brains and brawn.

Chapter 13

Common and Differentiated Entitlements: Pathway for Food Security for All

Pranab Mukherjee in his budget speech delivered on 26 February 2010 announced: "We are now ready with the draft Food Security Bill which will be placed in the public domain very soon". Although no official draft has so far been placed on the website of the concerned Ministry, several leading organisations and individuals have questioned the adequacy of the steps proposed to be taken under the Bill for achieving the goal of a hunger-free India. Based on Article 21 of the Constitution, the Supreme Court of India has rightly regarded the right to food as a fundamental requirement for the right to life. Many steps have been taken since Independence to adopt Mahatma Gandhi's advice for an *antyodaya* approach to hunger elimination. In spite of numerous measures and social safety net programmes, the number of undernourished persons has increased from about 210 million in 1990–1992 to 252 million in 2004–2006. About half of the world's under-nourished children are in India. Also, there has been a general decline in per capita calorie consumption in recent decades. Grain mountains and hungry millions continue to co-exist.

We are fortunately moving away from a patronage to a rights approach in areas relating to human development and well-being. The Acts relating to the Right to Information, Education, Land for Scheduled Tribes and Forest Dwellers, and Rural Employment are examples of this trend. The National Food Security Bill, when enacted, will become the most important step taken since 1947 in launching a frontal attack on poverty-induced endemic hunger in the country. The adverse impact of undernutrition on human health and productivity is well known. In addition to denying every citizen the right to a productive and healthy life, chronic undernutrition also makes it difficult to overcome diseases like tuberculosis, leprosy and HIV/AIDS. In the case of such diseases, a food-cum-drug approach is needed to ensure success of the treatment.

The numerous programmes introduced by the Government of India from time to time for improving the nutritional status of children, women and men include the following:

- *Ministry of Women and Child Development*: Integrated Child Development Services (ICDS), Kishori Shakti Yojana, Nutrition Programme for Adolescent Girls, Rajiv Gandhi Scheme for Empowerment of Adolescent Girls.
- *Ministry of Human Resource Development*: Mid-day Meals Scheme, Sarva Shiksha Abhiyan.
- *Ministry of Health and Family Welfare*: National Rural Health Mission, National Urban Health Mission.
- *Ministry of Agriculture*: Rashtriya Krishi Vikas Yojana, National Food Security Mission, National Horticulture Mission.
- *Ministry of Rural Development*: Rajiv Gandhi Drinking Water Mission, Total Sanitation Campaign, Swarna Jayanti Gram Swarojgar Yojana, Mahatma Gandhi National Rural Employment Guarantee Programme.
- *Ministry of Food*: Targeted Public Distribution System (PDS), Antyodaya Anna Yojana, Annapoorna.

Notwithstanding this impressive list of interventions, the present situation in the field of child nutrition is that 42.5 per cent of children below 5 years of age are underweight and 40 per cent of children below 3 years are undernourished.

In order to succeed in ensuring food security for all, we should be clear about the definition of the problem, the precise index of measuring impact and a road map for achieving the goal. Presently, the discussion is mainly around the definition of poverty and on methods of identifying the poor. We have the most austerely defined poverty line in the world and the official approach appears to be to restrict support only to Below Poverty Line (BPL) families. Calculations of the numbers of BPL families (taking 4 persons as the average size of a family), the number varies from 9.25 crore (Suresh Tendulkar Committee) to 20 crore (Justice DP Wadhwa).

Food security, as internationally understood, involves physical, economic and social access to a balanced diet, safe drinking water, environmental hygiene and primary healthcare. Such a definition will involve concurrent attention to the availability of food in the market, the ability to buy the needed food and the capability to absorb and utilise the food in the body. Thus, food and non-food factors (i.e., drinking water, environmental hygiene and primary healthcare) are involved in food security.

As earlier mentioned, we have numerous Central Government schemes dealing with nutrition support, drinking water, sanitation and healthcare. In addition, most State Governments have additional schemes such as extending support to mothers to feed new born babies with mothers' milk for at least the first six months. States like Tamil Nadu and Kerala have the universal PDS. Unfortunately, the governance of the delivery of such programmes is fragmented and a "deliver as one" approach is missing. Also, a lifecycle approach in the development and delivery of nutrition support programmes, starting with pregnant women and ending with old and infirm persons, is lacking. Our current unenviable status in the world in the field of nutrition is largely because of the absence of a good governance system which can measure outlay and output in an unbiased manner. Therefore, more than new schemes, the governance of existing schemes needs attention.

The Various Entitlements

In my view, the National Food Security Bill should be so structured that it provides common and differentiated entitlements. The common entitlements should be available to every citizen of the country. This should include a universal PDS, clean drinking water, sanitation, hygienic toilets, and primary health care. The differentiated entitlements could be restricted to those who are economically or physically handicapped. Such families can be provided with wheat or rice in the quantity decided at Rs. 3 per kg, as is now being proposed. The availability of cheap staple grains will only help BPL families address the problem of access to food at affordable prices, but not give them economic access to balanced diets. At the prevailing price of pulses, such families will find protein-rich foods out of reach. Similarly, hidden hunger caused by micro-nutrient deficiencies like iron, iodine, zinc, vitamin A and vitamin B12 will continue to persist. The question thus arises as to what we want to achieve from the proposed National Food Security Bill. Should the Bill enable every child, woman and man to have an opportunity for a healthy and productive life, or just have access to the calories required for existence? If the aim is the latter, the title "National Food Security Bill" will be inappropriate.

We can learn much from the "Zero Hunger" programme of Brazil, where a holistic view of food security has been adopted. The measures include steps to enhance the productivity of smallholdings as well as the consumption capacity of the poor. Our farmers will produce more if we are able to purchase more. Emphasis on agricultural production, particularly small farm productivity, will as a single step make the largest contribution to poverty eradication and hunger elimination. While universal PDS should be a legal entitlement, the other common entitlements could be indicated in the Bill for the purpose of monitoring and integrated delivery. The involvement of gram sabhas and nagarpalikas in monitoring the delivery systems will help to improve efficiency and curb corruption.

What is desirable should also be implementable. The greatest challenge in implementing the common and differentiated food entitlements under the Bill will be the production of adequate quantities of staple grains. Fortunately, the untapped production reservoir, even with the technologies now on the shelf, is high in both irrigated and rainfed farming systems. Doubling the production of rice and wheat in eastern India and pulses and oilseeds in rainfed areas during this decade is feasible. The 2010–2011 budget indicates measures for initiating a "bridge the yield gap movement" in eastern India, and for stimulating a pulses and oilseeds revolution through the organisation of 60,000 Pulses and Oilseed Villages where concurrent attention will be given to the conservation of soil and water, cultivation of the best available strains, consumption of local grains (like jowar, bajra, ragi, etc.) and commerce at prices fair to farmers. National and State efforts should be supported by efforts at the local body level for building a community food security system involving seed, grain and water banks.

The National Commission on Farmers (2006) in its comprehensive recommendations on building a sustainable national nutrition security system had calculated that about 60 million tonnes of foodgrains will be needed for universal PDS. The differentiated entitlements for BPL families for foodgrains at low cost will involve only additional cash expenditure. In fact, food stocks with Government may touch 60 million tonnes by June 2010.

For the Government to remain at the commanding height of such a food security system combining universal and unique entitlements, the four-pronged strategy indicated in Pranab Mukherjee's budget speech should be implemented jointly by Panchayats, State Governments and Union Ministries with speed and earnestness. Just as the Golden Quadrilateral initiative of Atal Behari Vajpaye electrified the national road communication infrastructure, we need a Golden Quadrilateral in the development of a national grid of modern grain storages. Will Manmohan Singh leave his footprints on the sands of time, in the case of safe storage of foodgrains and perishable commodities all over the country as an essential requirement for food security, as Vajpaye had done in the case of roads?

I hope we will not lose this historic opportunity for ensuring that our nation takes to a development pathway which regards the nutrition, health and well-being of every citizen as the primary purpose of a democratic system of governance.

Chapter 14

Designing Architecture for a Learning Revolution

Our performance on the economic front has been impressive in recent years with the growth in Gross Domestic Product (GDP) rising from 8.4 per cent in 2005–2006 to 9.2 per cent in 2006–2007. Unfortunately, such a creditable achievement has not been accompanied by equitable growth, with the result that divides like urban–rural, gender, economic, technological, and social divides like caste, tribe, religion and region are increasing. Let me take the case of nutrition. While the prevalence of clinical forms of protein energy malnutrition has decreased significantly, the sub-clinical forms such as underweight, stunting and wasting among children below five remain significantly high. About 23 per cent of newborns in India are of low birthweight due to maternal and foetal undernutrition and malnutrition. According to National Family Health Survey No. 3, about 43 per cent of under-five children are underweight and 48 per cent are stunted. About 36 per cent of adult women and 34 per cent of adult men suffer from chronic energy deficiency. Surveys carried out by the National Nutrition Monitoring Bureau (NNMB) during 2005–2006 in eight States revealed that about 49 per cent of 10–13 year-old girls and 18 per cent of 14–17 year-old adolescent girls in the rural areas are undernourished.

Our poverty line is one of the most austerely defined in the world, since it takes into account only the minimum food intake needed for survival. In spite of all the poverty alleviation programmes undertaken during the last 60 years, both rural and urban poverty remain high with 28.3 per cent in rural areas and 25.7 per cent in urban areas remaining below the poverty line. The percentage of illiterate women and men still remain high. According to the Census of India 2001, literacy rate ranged from 47 per cent in Bihar to 90.9 per cent in Kerala. Mizoram had also a high literacy rate of 88.8 per cent. Female literacy was lower in all the States. The position however has been improving since 2001, thanks to Sarva Shiksha Abhiyaan.

The great freedom poet Subramania Bharathi emphasised that food and education constitute the two legs of a human being and that it is the fundamental duty of independent India to ensure nutrition and education for all. It is obvious that in both these areas, we have a big deficit of achievement. With a net addition of about 17 million every year to our population, we have to run twice as fast to stay where we are, as emphasised frequently by Pandit Jawaharlal Nehru. How then can we achieve the goals of Sarva Shiksha Abhiyaan, Integrated Child Development Services (ICDS), Noon Meal Programme in Schools and other government programmes designed to address the issues of undernutrition and illiteracy in an integrated manner?

In the area of nutrition, it is now realised that there is need for a lifecycle approach. We have to start with pregnant women in order to ensure that healthy babies are born. How can we organise our education programmes on similar lines, starting with children, to foster a learning revolution based on the principle of social inclusion in opportunities for quality education?

There has been considerable work in our country on early child care and education.

> In the first year of life, the young human being learns more than he ever does in any other year of his life. This is the period of maximum learning and the rate of growth and development is phenomenally high. Starting from almost nothing, the child learns to move, walk and communicate within the short space of twelve months. Learning proceeds rapidly during the next few years, but the rate of growth gradually slows down. The first five years are the period of maximum intellectual development (80 per cent is completed by this time). Of this, 40 per cent is accomplished in the first year of

> life, and another 40 per cent by the end of the fourth year. These facts alone make this period one of great potential educational significance, and provide the strongest argument in favour of a strong base of education at the pre-primary level. (National Focus Group on Early Childhood Education)

The learning process is related to the self-activity of the child. Essential ingredients are a wide range of activities, free choice among them, the guidance of a skilled teacher, problem-solving situations and tasks graded to the ability and stage of development of the individual, a permissive environment, stimulus and challenge to exploration. However, the main idea is that learning is an active process and not a passive acceptance of all things told by the teacher. The corollary of this is that there is no such thing as teaching, there is only learning. Education involves totality. The child is learning all the time, not at set times. He is learning in a variety of areas, and with all aspects of himself. His intellectual, physical, emotional, aesthetic, social and moral learning proceed side by side.

Gurudev Tagore's poem in *Gitanjali* on what we can learn from the child captures the essence of child education.

> Child, how happy you are sitting in the dust
> Playing with a broken twig all the morning;
> I smile at your play with that little bit of broken twig
> I am busy with my accounts adding up figures by the hour
> Perhaps you glance at me and think what a stupid game to spoil your morning with
> Child, I have forgotten the art of being absorbed in sticks and mudpiles
> I seek out costly playthings and gather lumps of gold and silver
> With whatever you find, you create glad games;
> I spend both my time and my strength over things I can never obtain;
> In my frail canoe I struggle to cross the sea of desire;
> And forget that I too am playing a game ...

In spite of the importance of this area in the field of education, attention and support have not kept pace with need. Professor Krishna Kumar has pointed out that "in the context of early childhood education, even national level flagship programmes such as the ICDS have lost their claim to financial favour, what to say of smaller state-level initiatives. The hope of curricular reform in nursery education has dwindled even as aggressive private players have set up business, offering a cacophony of electronic devices and formal instruction through a frugally trained housewife doubling up as a nursery teacher". (*The Hindu*, 2 July 2008).

At the primary level, play gradually turns into work in two ways. First, the activity is undertaken with a definite objective in view. There is an aim to be achieved, and a task to be done. Secondly, the tasks are no longer freely selected by the child, but are socially determined and in practice set by the teacher. Nevertheless, the same purpose can be achieved as in his play, if the task is seen by the child to have some social and personal significance. A socially productive and useful task can be an agent of the education.

It is a tribute to the genius and vision of Mahatma Gandhi that he identified and discussed at length nearly 80 years ago methods of making our entire population literate. Gandhi's contribution to educational thought was unique and typical of him. The wonder of his approach is that, by merely applying himself to the problems of Indian education as it was and as it should be, and out of the depths of his intuitive understanding and judgment, he arrived at the same conclusions as those which the best educational thinkers have evolved during the early part of the 20th century. I wonder whether he was acquainted, except in a casual manner, with the writings of Rousseau, Froebel, Pestalozzi, Dewey or Maria Montessori. Yet the principles behind what later came to be known as Basic Education are identical to those stressed by all these educationists, i.e., the principle of activity as the basis of learning. Gandhi's concept of Basic Education was essentially that of activity education — it was both craft-centered and child-centered and was thus diametrically opposed to the academic and book-centered system. Dr. Zakir Hussain attempted to carry this movement forward but sadly had to concede that "Basic Education in our country has been a failure, not because it was educationally unsound but because it was never given a fair trial... We have turned the so-called intellectual book school into a mechanical memory training school".

With simple tools such as a soil testing kit and seed kit, a whole new world can be opened up for the school children in villages. The study of birds, the identification of weeds, the detection of alkalinity, the harvesting of rainwater and the prevention of damage by rats and pests both in the field and in the storerooms would all have immense educational and practical value. The equipment needed for such studies is simple and inexpensive and mostly requires only a well-informed teacher who does not curb the questioning mind and is not afraid of long walks. With a little training this is one field where all graduate students of agriculture and science can render great service. This experience will enrich their own understanding of our biological assets and problems and at the same time make Mahatma Gandhi's and Dr. Zakir Hussain's dream of a learning revolution come true.

In his fascinating book *Last Child in the Woods* (2006), Richard Louv has drawn attention to the links between the absence of nature in the lives of today's wired generation and some of the most disturbing childhood trends such as the rise in obesity, attention disorder and depression. He has termed this phenomenon nature-deficit disorder. By weighing the consequences of the disorder, we can also become aware of how blessed our children can be — biologically, cognitively, and spiritually — through positive physical connection to nature.

Turning now to university education, the Education Commission (1964–1966) led by Professor DS Kothari offered several practical suggestions to make university students "employable", or capable of self-employment. In agriculture, the Commission recommended the establishment of polytechnics where students could acquire the necessary technical skills so as to contribute towards the growth of scientific agriculture. The Commission also suggested ways of achieving a qualitative change in education and of linking admissions to the country's ability to absorb university-trained personnel. Many of the Commission's suggestions are yet to receive serious consideration from the view point of implementation. Meanwhile, the problem of jobs and student unrest is growing and there is migration of both educated and uneducated youth from villages to towns in search of work opportunities. Our education, while failing to increase our wealth-creating capacity, has enhanced our wealth-absorbing desire.

With the spread of new technologies and the dramatic transformation of agriculture in certain areas, new dimensions of adult education also appear. There are new needs for education among farming communities. There is a great hunger not only for new knowledge related to agriculture but also for new skills, particularly technical skills connected with it. The demand for "techniracy", a term I coined for defining the pedagogy of learning the latest technical skills through work experience, is likely to be much stronger and deeper and also more widespread than that for formal literacy, or even for functional literacy. New approaches to adult education must capitalise on this new demand and need for techniracy. I later developed the concept of Krishi Vigyan Kendras (KVKs) to provide an institutional mechanism for imparting techniracy. KVKs now exist in almost all districts of the country. But, they are unfortunately tending to become routine institutions with farm women and men remaining passive recipients of information.

In addition to KVKs, there is need for establishing Farm Schools in outstanding farmers' fields. This will help to promote farmer-to-farmer learning which has high credibility because of the trust imposed by farmers on the economics of the farm enterprise of fellow farmers. Also, KVKs can have as staff members and resource persons practising farmers who know the art and science of farming from field experience. It is said that, "one ounce of practice is worth tonnes of theory". This is very true in farming where real life experience is the best guide for sustainable advances in productivity. Today we can see that techniracy can be achieved through Information Communication Technology (ICT) in rural areas on topics vital to the livelihoods of the rural masses.

It is common knowledge that women graduates often get most of the academic awards in convocations. Unfortunately, opportunities for the full expression of their intellectual brilliance and academic commitment become fewer once they leave the university. The multiple burdens on their time, such as child rearing, home keeping and economic activities give them inadequate time to continue with their preferred professional pursuits. It would be useful if the placement bureaus of universities have a special wing for assisting graduates who need opportunities for self-employment characterised by flexible timings in terms of working hours. We should not lose the services of large numbers of women graduates with outstanding academic records due to gender insensitivity in academic and personnel policies.

In the academic world, the era of working for others has to give way to the era of working with others. It would be useful to recall what Gurudev Tagore said in this context: "A candle which is not lit cannot light others; a teacher who is also not learning cannot teach others". A good teacher thus takes the role of a life-long student.

Both the need and the opportunities for a "learning revolution" have never been as great as they are now. What is needed is the will to act and not mere discussion and analysis. "For sheer size, the tasks ahead of us are so demanding that no one can afford to sit back and just watch or let frustration become endemic in our country. The situation demands of us work, work and more work, silent and sincere work, solid and steady reconstruction of the whole material and cultural life of our people" — these words are even more relevant today then when Dr. Zakir Hussain spoke them while assuming the office of the President of India on 13 May 1967.

Chapter 15

Role of Sustainability Science

The Indira Gandhi National Open University (IGNOU) has a very wide reach and nearly two million students all over the country are benefiting from this unique centre for distance education. IGNOU also adopts a dynamic approach in developing new courses and curricula. For example, I now serve as its Honorary Chair for Sustainable Development.

Sustainable Development is based on the foundation of sustainability science supported by a multi-disciplinary approach. Such an inter-disciplinary science has to be built on the following guiding principles.

Ethics: Ethical considerations will have to guide human behaviour in relation to natural resources exploitation. Bioethics and environmental ethics are now developing into well-defined scientific areas. The ethical responsibility of safeguarding the environment rests on professionals, political leaders and the public. In the past, by investing conservation with spiritual significance, every individual was made to integrate ethics in day-to-day life. Advances in molecular genetics and biotechnology, which permit us to "play God", have increased the urgency of bioethics courses in our universities.

Economics: Ecological economics does not permit depreciation of natural assets. Thus it has a time dimension of infinity. Ecological economics is also a fast-developing science and it will help to measure the benefit-risk structure of development projects from the point of view of their long-term impact. Ecological economics should become part of the curriculum in technological and management institutions. All dependent on natural resources for their enterprises should understand that good ecology is the pathway to good and enduring business.

Equity: The concept of equity is now discussed in terms of both intra-generational equity and inter-generational equity (i.e., safeguarding the interests of the future generations). For example, over-exploitation and pollution of the aquifer will deny opportunities for groundwater availability to the generations yet to be born. Similarly, the melting of ice and glaciers resulting in water shortage in cold desert areas like Ladakh will force future generations to migrate from the area. Climate change leading to the melting of ice will not only cause floods in the plains but also a rise in sea level over a period of time. Another important component of equity relates to the gender dimension of sustainability science. Women have been great conservers of biodiversity and natural resources. Their role should be acknowledged and strengthened.

Energy: Energy is a key factor in terms of both economic development and climate change. Integrated energy supply systems involving the optimum use of all renewable forms of energy like solar, wind, biomass, biogas and geothermal have to be developed. Other opportunities like hydrogen and nuclear energy will have to be integrated into an overall sustainable energy security system.

Employment: Many of the livelihood opportunities in developing countries are based on the use of natural resources like land, water, forest and biodiversity. Emerging technologies tend to promote jobless economic growth. In population-rich but land- and water-hungry countries, there is a need for job-led economic growth. Therefore development experts and technology developers should take into account the impact of new technologies and management procedures for work and livelihood security. Jobless growth is joyless growth in population-rich countries like ours.

Education: Education is a cross-cutting theme and has to take into account all the above-mentioned factors. Environmental literacy should be based on the principle of "do ecology". For example, in the case of biodiversity, there is a need to create an economic stake in conservation. "Orphan" crops can be saved only if there are markets for them. Similarly, in the case of nature tourism, those who operate houseboats or hotels in eco-sensitive areas should be made aware that good ecology is good business. Environmental education should also be based on practical examples which can drive home the message which is to be conveyed. Therefore, it should be based on field projects which can demonstrate how to organise ecotourism, conduct green audit or manage rainforests sustainably. Just as action research programmes help to gather data on the economics and ecology of development projects, action education will derive its roots from field experience.

Countries like ours require "do" ecology and not just "don't" ecology. Education should go to the grass-roots level and in this respect India is fortunate to have grass-roots democratic institutions like panchayats and nagarpalikas. Elected members of these bodies should become environmentally literate. This is where modern Information Communication Technology (ICT) involving the integrated use of the internet, cable TV, community radio and cell phone will help to achieve the last mile and last person connectivity in terms of knowledge empowerment. Distance education methods as promoted by IGNOU can help to reach the unreached and voice the voiceless.

Population growth should not exceed the population-supporting capacity of ecosystems. The human ecological footprint should be reduced through limiting wants and avoiding waste. Many years ago, Gandhiji said: "Nature provides enough for everyone's need, but not for anyone's greed." Yet, today over a billion women, men and children of the human population are living in absolute poverty and destitution, while another billion are leading unsustainable lifestyles. Therefore, the ethical principles propagated by sustainability science should aim to curtail both poverty and unsustainable consumption of natural resources. This is the challenge before us from the point of view of ensuring the well-being of both the present and future generations.

Sustainability science is both multidisciplinary and multidimensional. For each area of human activity, there is a need to develop technologies which can help to achieve the desired goal without associated ecological harm. For example, in the case of agriculture which occupies the largest land area and utilises over 75 per cent of water resources, there is a need for developing methodologies to achieve an Evergreen Revolution which can ensure enhancement of productivity in perpetuity without associated ecological harm. Conservation farming and green agriculture which involve the use of integrated natural resources and pest management techniques are the pathways to an evergreen revolution. Sustainability science involves both anticipatory research, as for example in the case of meeting the challenges of climate change, as well as participatory research and knowledge management with rural and tribal communities in order to ensure that the recommended practices are socially compatible and economically feasible. Also, education has to be derived from the adoption of an agro-climatic and agro-ecosystem approach, taking into consideration the specific needs and opportunities prevailing in arid, semi-arid, hill, coastal, irrigated and island ecosystems.

Harmony with nature should become a non-negotiable ethic. The rise and fall of great civilisations in the past have been related to the use and abuse of land, water and other natural resources. Therefore, sustainability science should guide all technology development and dissemination programmes of our universities and research institutions.

Chapter 16

Towards Eliminating Hunger and Poverty

In September 2010, a Summit was held in New York under the auspices of the United Nations to review the progress made during the last ten years in achieving the targets set under the UN Millennium Development Goals (MDGs) adopted by Member Nations in 2000. The UN-MDGs represent a global common minimum programme for sustainable human security and well-being. In spite of the modesty of the goals set, progress in achieving them has been inadequate in many developing countries, including India. In fact, the Food and Agriculture Organization (FAO) points out that the number of children, women and men going to bed hungry now is over a billion, although this number was only 800 million in the year 2000. There is obviously a need to review our strategies and redouble our efforts in achieving all the UN-MDGs and particularly the very first one relating to halving hunger and poverty by the year 2015.

The Economic Survey of India (2008) contained the following observations:

- While poverty rates have declined significantly, malnutrition has remained stubbornly high. Malnutrition, as measured by underweight children below 3 years, constitutes 45.9 per cent as per the National Family Health Survey 2005–2006 (NFHS 3). It has also not significantly declined from its level of 47 per cent in 1998–1999 (NFHS 2).

- It is evident that existing policies and programmes are not making a significant dent on malnutrition and need to be modified. While per capita consumption of cereals has declined, the share of non-cereals in food consumption has not grown to compensate for the decline in cereal availability.

For achieving sustainable food security, concurrent attention will be necessary to ensure food availability, access and absorption. Access depends upon opportunities for employment, while absorption will be conditioned by clean drinking water, sanitation and healthcare. Thus, both food and non-food factors regulate food security.

In the case of the economically underprivileged sections of our population, over 70 per cent of their income goes to the purchase of food. This is why there is a strong interconnection between hunger and poverty. India used to witness serious famines during the colonial rule. The last big famine was the Bengal Famine of 1943. Although famines of this order have not occurred in independent India, chronic hunger arising from inadequate purchasing power and hidden hunger caused by micronutrient deficiencies in the diet like iron, iodine, zinc, vitamin A, vitamin B12 etc., are widespread. What is the role of science in dealing with such issues?

A major development in Indian agricultural science was the setting up of the Indian Agricultural Research Institute, the Indian Veterinary Research Institute, and the National Dairy Research Institute during the later part of the 19th century and the beginning of the 20th century. Later, the Indian Council of Agricultural Research (ICAR) was established on the recommendations of the Royal Commission on Agriculture headed by Lord Linlithgow. The Commission said:

> However efficient the organisation which is built up for demonstration and propaganda be, unless that organisation is based on the solid foundation provided by research, it will be merely a house built on sand.

We now have a fairly impressive infrastructure in respect of agricultural research and education. In addition to ICAR institutions, there are a large number of agriculture, animal sciences, fisheries and horticulture universities. Still poverty and hunger persists at inexcusable levels. It is under these conditions that the MS Swaminathan Research Foundation (MSSRF) was established in 1988. I would like to take the work being done under six major interdisciplinary areas at MSSRF to illustrate the kind of reorientation we need in our research and educational strategies.

Coastal Systems Research: First, I would like to emphasise the need for Coastal Systems Research (CSR). Globally, nearly 97 per cent of water is seawater. Unlike farming systems research which deals with crop-livestock production systems, the coastal system which needs integrated attention to the landward and seaward sides of the coast is yet to receive similar attention. This is why MSSRF chose CSR as a priority area. India has nearly 7500 km of coastal area in addition to the Andaman and Nicobar and Lakshadweep group of islands. There is hence enormous potential for integrated agri-aqua systems of farming. This will involve the cultivation of halophytes including mangroves and the culture of salt water-tolerant fishes. The mangroves and other halophytes also serve as bioshields, performing the role of speed breakers during cyclonic storms and tsunamis. The year 2010 is the 80th anniversary of the Dandi Salt March led by Mahatma Gandhi to emphasise that seawater was a social resource and should not be taxed by government. Unfortunately, seawater farming based on coastal agro-forestry and capture and culture fisheries has not yet become widespread. Nearly 150 years ago, farmers in areas like Kuttanad in Kerala had developed techniques for below sea level farming. This knowledge will be very useful in the context of a potential rise in sea level. Nearly 25 per cent of our populations live within 50 km of the shoreline. Global warming is likely to result in a rise in sea level within the next few decades. Therefore, we should launch an integrated coastal zone management strategy linking the ecological security of coastal areas and the livelihood security of coastal communities in a mutually reinforcing manner. Such a strategy should include both seawater farming through agri-aqua farms and below sea level farming based on the Kuttanad experience.

Biodiversity: 2010 is being observed as the International Year of Biodiversity in order to highlight the critical role biodiversity plays in the areas of food, health, livelihood and environmental security. Agro-biodiversity rich areas are also rich in cultural diversity. MSSRF's work in Kolli Hills in Tamil Nadu, Wayanad in Kerala and Koraput in Odisha has shown that tribal families, who observe a strict conservation ethic, tend to remain poor. We must break the nexus between the poverty of the people and the prosperity of nature if we are to safeguard biodiversity for future generations. MSSRF's approach has been to link commercialisation and conservation in a mutually reinforcing manner so that there is an economic stake in conservation. In June 2010, delegates from 90 countries, meeting in Busan, Republic of Korea, approved the establishment of an Intergovernmental Platform on Biodiversity and Ecosystem Services (IPBES) on the model of the Inter-Governmental Panel on Climate Change (IPCC). It will be prudent to set up a National Platform on Biodiversity and Ecosystem Services in order to generate synergy among ongoing programmes.

Biotechnology: Recombinant Deoxyribonucleic acid (DNA) technology has provided powerful tools for moving genes across sexual barriers and for developing novel genetic combinations. It is important to use this tool for solving present and potential problems arising from unfavourable temperature, rainfall and sea level. Therefore, during the last twenty years, MSSRF scientists have concentrated on identifying genes for salt water and drought tolerance. A US patent has been granted for the dehydrin gene from *Avicennia marina* responsible for salt tolerance in plants. Similarly Glutathione S Transferase (GST) gene from *Prosopis juliflora* conferring resistant to drought has also been granted a US patent. These are very valuable genes and have to be combined with crop varieties having desirable agronomic and culinary characteristics. This has already been done by MSSRF scientists. Another area where recombinant DNA technology can be useful is in bio-fortification. Iron (Ferritin) rich rice varieties have been developed using genes from *Avicennia marina*. Thus there are uncommon opportunities for developing climate-resilient strains of crop plants, farm animals and fishes. Genes like Sub-1 in rice provide apportunities for breeding varieties for flood tolerance. There is need for setting up Gene Banks for a warming India.

Ecotechnology: Knowledge is a continuum. We cannot place traditional and modern knowledge into two different pigeonholes. Modern knowledge has its roots in ancient wisdom. Ecotechnology helps to blend traditional ecological prudence and techniques with frontier science and technology. Ecotechnology gives concurrent attention to ecology, economics, ethics, equity, energy and employment generation. A method of converting ecotechnology into jobs and income is through biovillages, where simultaneous attention is given to natural resources conservation and enhancement, improvement of small farm productivity and profitability, and generation of non-farm employment. A Rural Systems Research (RSR) methodology helps to improve rural professions — farm and non-farm — in an integrated manner. The National Policy for Farmers placed in Parliament in November 2007 calls for as much emphasis on farmers' income as on production. Such an income-orientation to farming can be achieved only through RSR. Unfortunately, agricultural universities and research institutions are yet to adopt such an integrated approach to improving agrarian and rural prosperity.

Food Security: In recent years there has been a paradigm shift from a patronage to a rights approach in relation to information, education, employment and, in the case of tribal families, ownership of land. The Government of India has committed to bring food security also under the category of legal right. A sustainable food security system will depend on adequate production, procurement on the basis of a minimum support price, preservation in modern silos or other forms of storage, and, above all, an efficient and corruption-free public distribution system. The National Food Security Act, when enacted, provides a great opportunity for stimulating the conservation of natural resources, cultivation using new technologies, consumption of a wide range of grains, and farmer-centric marketing. While right to information can be enforced through files, the right to food has to come from the farmer and the field. The right to food can be maintained only if there is increase in productivity in perpetuity without ecological harm, i.e., the evergreen revolution, spearheaded by families with smallholdings. Our agriculture is at the crossroads. Farm ecology and economics are getting adverse to sustained productivity. Inadequate public good research further compounds the problem of technology choice and access. We should ensure that the National Food Security Act covers a wide range of staples and not just wheat and rice. At the same time, there is a need to strengthen the ecological foundations of sustainable agriculture, particularly with reference to land, water, biodiversity and climate.

Information, Communication Technology (ICT): Bridging the urban–rural digital divide helps to bridge economic, skill and gender divides. Biotechnology, space technology and ICT are transformational technologies. We should make every village a knowledge centre in order to take the benefits of modern scientific knowledge and techniques to rural professions. Mahatma Gandhi urged that there should be a marriage between brain and brawn if Indian agriculture is to progress. This can be achieved through the effective use of ICT based on location-specific needs and language. The Grameen Gyan Abhiyan provides a great opportunity for taking the benefits of ICT to the rural poor, based on a last mile and last person connectivity. Synergy between the internet and cell phone or FM radio and cell phone helps to take the benefits of right information to the right place at the right time. A rural knowledge revolution is vital for ending all forms of divides and substituting them with the technological and skill upgradation of rural professions.

Way Forward

- There is a need to revamp and refocus agricultural research and education in order to achieve the paradigm shift from assessing progress only in terms of millions of tonnes of grains to measuring it in terms of the growth in the real income of farm families, as envisaged in the National Policy for Farmers presented to Parliament in November 2007.

- A decentralised food security system involving the setting up of village-level Grain Banks will be very effective.
- CSR, RSR and farmer-participatory research and knowledge management are important for increasing the income and employment potential of coastal and rural professions on an environmentally sustainable basis.
- Strategic research involving marker-assisted selection or genetic engineering techniques will be necessary for developing farming practices which can minimise the adverse impact of climate change.
- Lab to Land can be facilitated through information and communication technologies based on an appropriate mix of the print media, cable TV, internet, FM radio and the mobile phone.
- To help rural families to adapt to climate change, 127 Research and Training Centres should be established, one in each of the 127 agro-climatic subzones. These Centres should develop computer simulation models of alternative cropping strategies to suit different weather models. Seed Banks should be established at these centres to provide seeds of the alternative crops. Safe grain storage structures should also be established in each of the 127 agro-climatic zones. Our whole aim should be to maximise the benefits of a good monsoon and minimise the adverse impact of unfavourable weather.
- A national agricultural research agenda for meeting the challenge of climate change is an urgent need. In wheat, we will have to change the focus from per crop yield to per-day productivity. In potato, we should develop True Potato Seed (TPS) technology, in case seed tubers are affected by virus diseases due to vector activity. In rice, we should develop strains containing Sub-1 gene for imparting flood tolerance.

The future of our food security system will depend upon the scientific and policy support we extend to our farming community, who constitute one-fourth of the global farming community. The

Green Revolution was the result of a small government programme getting converted into a mass movement led by farm men and women. Today, our educational and research institutions are more obsessed with bricks rather than with brains. We must reverse the paradigm and nurture brains which can help to promote knowledge-intensive agriculture.

Section III

Food Security in an Era of Climate Change and Civil Strife

Chapter 17

Copenhagen, Tsunami and Hunger

The principle of common but differentiated responsibilities is the core of the many climate agreements arrived at so far, including the Kyoto Protocol (1997) and Bali Plan of Action (2007). The differentiated responsibilities aim to meet the special needs of developing countries for accelerated and equitable economic development. Both at L'Aquilla and Copenhagen, the industrialised countries proposed limiting the rise in mean temperature to 2°C above normal. Even this seems to be unattainable in the context of the present rate of emission of greenhouse gases (GHGs). Hence, the principle of common but differentiated impact of a 2°C change in mean temperature is essential for prioritising climate victims. For example, small islands like Tuvalu in the Pacific Ocean, the Maldives, Lakshadweep and Andaman and Nicobar, as well as Sunderbans in West Bengal, Kuttanad in Kerala and many locations along the coast will all face the prospect of submergence. Floods will become more serious and frequent in the Indo-Gangetic plains. Drought-induced food and water scarcity will become more acute. South Asia, Sub-Saharan Africa and the small islands will be the worst victims. In contrast, countries in the northern latitudes will benefit due to longer growing seasons and higher yields.

Addressing the World Climate Conference held in Geneva in 1989 on the theme "Climate Change and Agriculture", I pointed out the serious implications of a rise of 1 to 2°C in mean temperature on crop productivity in South Asia and Sub-Saharan Africa. An expert team constituted by FAO, in its report submitted in September 2009, also concluded that for each 1°C rise in mean temperature, wheat yield losses in India are likely to be around 6 million tonnes per year, or around US$1.5 billion at current prices. There will be similar losses in other crops and our impoverished farmers could lose the equivalent of over US$20 billion in income each year. Rural women will suffer more since they look after animals, fodder, feed and water.

We are in the midst of a steep rise in the price of essential food items like pulses. 2009 has been characterised both by extensive drought and severe floods. The gap between demand and supply is high in pulses, oilseeds, sugar and several vegetable crops including onion and potato. The absence of a farmer-centric market system aggravates both food inflation and rural poverty. The Food Agriculture Organisation (FAO) estimates that a primary cause for the increase in the number of hungry persons, now exceeding over a billion, is the high cost of basic staples. As I have repeatedly lamented, India has unfortunately the unenviable reputation of being the home for the largest number of undernourished children, women and men in the world. The task of ensuring food security will be quite formidable in an era of increasing climate risks and diminishing farm productivity.

China which was reluctant in Copenhagen to join other developing countries in efforts to restrict the rise in mean temperature to 1 to 1.5°C, has already built strong defences against the adverse impact of climate change. During this year, China produced over 500 million tonnes of foodgrains in a cultivated area similar to that of India. Chinese farm land is however mostly irrigated unlike in our country where 60 per cent of the area still remains rain-fed. Food and drinking water are the first among our hierarchical needs. Hence while assessing the common and differentiated impact of a 2°C rise in temperature, priority should go to agriculture and rural livelihoods.

What are the steps we should take in the field of both mitigation and adaptation?

The largest opportunity in the area of mitigation lies in increasing soil carbon sequestration and for building up soil carbon banks.

Increase in the soil carbon pool in the root zone by 1 tonne C/ha/yr will help to increase food production substantially, since one of the major deficiencies in soil health is low soil organic matter content. There should be a movement for planting a billion "fertiliser trees" which can simultaneously sequester carbon and enhance soil nutrient status. We can also contribute to the reduction in methane emission in the atmosphere from animal husbandry by spreading biogas plants. A biogas plant and a farm pond in every farm will make a substantial contribution to both reducing GHG emission and ensuring energy and water security. Similarly neem-coated urea will help to reduce ammonia volatilisation and thereby the release of nitrous oxide into the atmosphere.

2010 is the International Year of Biodiversity. We can classify our crops into those which are climate-resilient and those which are climate-sensitive. For example, wheat is a climate-sensitive crop, while rice shows a wide range of adaptation in terms of growing conditions. We will have problems with reference to crops like potato since a higher temperature will render raising disease-free seed potatoes in the plains of northwest India difficult. We will have to shift to cultivating potato from true sexual seed. The relative importance of different diseases and pests will get altered. The wheat crop may suffer more from stem rust which normally remains important only in peninsular India. A search for new genes conferring climate resilience is therefore urgent.

Anticipatory analysis and action hold the key to climate risk management. The major components of an action plan for achieving a climate-resilient national food security system will be the following:

- Establish in each of the 127 agro-climatic sub-zones, identified by the Indian Council of Agricultural Research (ICAR) based on cropping systems and weather patterns of the country, a climate risk management research and extension centre. Such centres should be manned by climate risk managers, who will blend traditional wisdom with modern science.
- Organise a content consortium for each centre consisting of experts in different fields to provide guidance on alternative cropping

patterns, contingency plans and compensatory production programmes, when the area witnesses natural calamities like drought, flood, higher temperature and in case of coastal areas, a rise in sea level.

- Establish with the help of the Indian Space Research Organisation (ISRO) a Village Resource Centre (VRC) with satellite connection at each of the 127 locations.
- Establish with the help of the Ministry of Earth Sciences and the India Meteorological Department an agro-meteorological station at each research and extension centre to initiate a "weather information for all" programme.
- Organise seed and grain banks based on computer simulation models of different weather probabilities and their impact on the normal crops and crop seasons of the area.
- Develop drought and flood codes indicating the anticipatory steps necessary to adapt to the impact of global warming.
- Strengthen coastal defences against rise in sea level as well as the more frequent occurrence of storms and tsunamis through the establishment of bio-shields of mangroves and non-mangrove species. Also, develop seawater farming and below sea level farming techniques. Establish major research centres for seawater farming and below sea-level farming. Kuttanad in Kerala and the Sunderbans area of West Bengal will be ideal locations for such centres.

26 December 2009 marked the fifth anniversary of the devastating 2004 tsunami, which was a wake-up call for alerting us to the consequences of a sudden rise in sea level. "Copenhagen inaction" will lead to more severe coastal storms, tsunamis and sea level rise. In 2010, India will be completing 60 years of planned development. Hereafter, climate resilience must be mainstreamed in all development programmes. Let not the Copenhagen inaction add to the number affected by deprivation and malnutrition.

Chapter 18

Monsoon Management in an Era of Climate Change

Mastering the art and science of monsoon management holds the key to adaptation to adverse changes in temperature and precipitation caused by climate change leading to the more frequent occurrence of drought and floods. Media accounts of the monsoon behaviour vary from day to day, from agony to ecstasy and then again to agony. Whether or not such deviations are related to climate change, the preparedness needed to enhance our coping capacity to meet the challenge of a very variable rainfall pattern is at present limited.

The southwest monsoon period (June–September), when we receive over 80 per cent of our annual rainfall, is not only important for our food and water security, but also for the work and income security of over 60 per cent of our population engaged in farming. The health and survival of nearly 500 million farm animals, including cows, buffaloes, goats and sheep also depend upon the availability of water for direct consumption and for the production of feed and fodder. In addition, power generation and the recharge of the aquifer are affected. Climate models predict that sub-regional variation in more extreme rainfall pattern resulting in drought or floods may increase. Experience indicates that some of the areas experiencing drought earlier in the year may face floods in August–September, resulting in a further damage to crops. Heavy rainfall and floods in August–September will however help to fill up tanks and reservoirs and augment groundwater supply, which will be very helpful for raising good *rabi* season crops.

In the case of monsoon rains, it is not only the total rainfall which is important, but more importantly, the distribution of the rainfall which will determine the health of the crop. This is an important problem we face in our country, where most of the rainfall occurs in about 100 hours in a year. The loss occurring from a skewed distribution of rainfall is particularly high in soils with limited water holding capacity.

Another aspect of climate change is the possibility of a rise in sea level and more frequent occurrence of coastal storms and cyclones. We should take anticipatory action to protect both the ecological security of coastal areas and the livelihood security of coastal communities. Preparing for climate change therefore should become an integral part of the planning process for sustainable food, water and livelihood security.

While the heartland of the Green Revolution experienced severe meteorological distress in 2009, other parts of the country had a reasonably satisfactory monsoon season, particularly eastern India comprising Bihar, Jharkhand, Chattisgarh, Odisha, West Bengal and Assam. Fortunately, the whole of eastern India as well as parts of peninsular India have a very large untapped yield reservoir even with the currently available technologies. Since there is enough soil moisture for raising good crops in the southern and eastern parts of the country, we should launch a compensatory production programme in this region. Such a programme should involve the intensification of the extension efforts, as well as supplying an additional dose of nutrients free of cost. It is important that the needed macro- and micro-nutrients are supplied in the form of top dressing of standing crops. The addition of 20 to 25 kg of nitrogen per hectare will help to increase yield by a tonne. The loss in yield sustained in the northwest part of the country can to some extent be made good by increasing production in those parts of the country which have experienced near normal monsoon conditions. If such a compensatory programme can be initiated by mobilising the requisite nutrients, it will be good if the faculty and scholars of Agricultural and Animal

Sciences Universities in the region stopped classroom education for a month, and went to villages to work with farm families in improving water and nutrient use efficiency and in controlling pests and diseases. I am confident that if the Farm Universities come to the rescue of the national food and livelihood security system by working with farm women and men for a month, they will also gain by learning more about the field problems than they will be able to do in the classrooms. With a National Food Security Act on the anvil, it will be prudent to develop procedures which can help us to optimise production in the most favourable areas from the point of view of moisture availability.

Monsoon irregularity has multiple adverse effects on crops, farm animals and human food and livelihood security in the most seriously affected areas. Also, hydropower generation is affected, leading to energy shortage. Power shortage in turn makes it difficult to give life-saving irrigation to crops, wherever opportunities for this exist. In such conditions, distress sale of farm animals is a clear index of extreme despair. Farm animal camps should be set up near a water source or near a groundwater sanctuary (i.e., concealed aquifer which can be exploited during emergency) where animals can be fed with agricultural residues enriched with urea and molasses.

Another urgent need is the launching "A Pond in Every Farm" movement, by permitting the Mahatma Gandhi National Rural Employment Guarantee Act (MGNREGA) workers to build *jal kunds* in the farms of small and marginal farmers. The revised MGNREGA guidelines permit this. At least 5 cents in every acre should be reserved for the construction of ponds to store rainwater. Where there is adequate groundwater, subsidised electricity and diesel should be made available on a priority basis. Energy is the key limiting factor in taking advantage of groundwater.

Over 60 per cent of our cultivated area is rain-fed. Considerable research has been carried out on increasing the yield and stability of crops in dry farming areas. The gap between potential and actual yields in these areas is as high as 200 to 300 per cent. There is need for launching a "bridge the yield gap movement" in all rain-fed areas through attention to varietal choice, soil fertility, water conservation and management and plant protection. Fortunately, the Rashtriya Krishi Vikas Yojana, involving an outlay of Rs. 25,000 crore, provides an opportunity for State Governments to help farmers to increase the production of dryland crops. Pulses and oilseeds are mostly cultivated with rainfall and we should increase the production of these crops to minimise imports and enhance the income of farmers.

There is also need for a decentralised grid of warehouses and grain storage structures based on recent technologies. This will help to prevent both distress sale and panic purchase. The production of vegetables and fruits is going up and it is important to strengthen facilities for the safe storage of perishable commodities.

Far back in 1972, while delivering the Dr. Rajendra Prasad Memorial Lecture, under the auspices of the Indian Society of Agricultural Statistics, on the topic "Can we face severe drought again without food imports?", I had stressed that just as grain reserves are important for food security, seed reserves are important for crop security. This is because the implementation of contingency plans involving the cultivation of crops based on the rainfall pattern will be possible only if the seeds of alternative crops are available. Later in 1973, delivering the Sardar Patel Memorial Lectures over All India Radio, I had underlined the need for preparing Drought, Flood and Good Weather Codes for each agro-climatic region in our country. The aim of such Codes, detailing the anticipatory measures that should be undertaken, is to minimise the adverse impact of aberrant weather and to maximise the benefits of good monsoons. Such action codes should become integral parts of our preparedness for insulating to the maximum possible extent our food, water and livelihood security systems from the impact of global warming.

An agreement was reached at the 2009 G8 meeting, attended by our Prime Minister, that greenhouse gas emissions should be reduced by 40 per cent by 2020, so as to limit the maximum rise in temperature to 2°C over the pre-industrial period. Even such a 2°C rise in mean temperature will have serious implications for our agriculture. Over 20 years ago, the late Dr. S.K. Sinha and I showed that a 1°C rise in night temperature in the Punjab and Haryana will reduce the duration of the wheat crop by about a week. This in turn will reduce yield by 4 to 5 quintals per hectare, resulting in a considerable loss in wheat production in that region.

Unlike in industrialised countries, the average size of our farms is hardly 1 ha. The smaller the farm, the greater is the need for marketable surplus to get cash income. Therefore, the implications of reduced yield should not be considered just in terms of production of wheat or rice or other crops, but also in terms of the income of the small farmer. To keep farmers in farming, there will be need for proactive action in enlarging rural livelihood opportunities in the non-farm sector.

At the same time, we should intensify research on assembling genetic material for a warming India. Novel genetic combinations for tolerance to higher temperature and moisture stress can be developed through the tools of recombinant Deoxyribonucleic acid (DNA) technology. Similarly, flood and salinity tolerant strains of major crops can be developed. Farming has to become knowledge and technology intensive if we are to successfully overcome the emerging challenges.

Chapter 19

Media and the Farm Sector

A major trigger for the Green Revolution, a term coined by Dr. William Gaud of the US Department of Agriculture in 1968 to mark a significant increase in crop production through yield advance, was the enormous enthusiasm generated among farm families by the print media and All India Radio on the opportunity created by semi-dwarf varieties of wheat and rice to enhance yield and income very substantially. The revolution resulted from a symphony approach consisting of four major components: technology, which is the prime mover of change; services, which can take the technology to all farmers whether small or large; public policies relating to the price of inputs and output; and above all, farmers' enthusiasm promoted by the mass media. I recall when we started large-scale research and testing with semi-dwarf varieties of wheat obtained from Mexico through Dr. Norman Borlaug in 1963, the new plant types attracted media attention immediately. Several enthusiastic reports appeared in our newspapers as well as in foreign journals like *The Economist* on the new opportunities opened up by scientists for achieving a quantum jump in yield. Such reports were based on the visits of experienced correspondents to the experimental fields of the Indian Agricultural Research Institute, New Delhi, and the Punjab Agricultural University, Ludhiana.

The media reports led to a widespread demand for the seeds of the new strains. To meet this demand, 18,000 tonnes of seeds of a few good varieties were imported in 1966 from Mexico, as a part of the purchase of time operation we had then designed. From 1964 to 1967, the country had the good fortune of having C. Subramaniam as Minister for Agriculture and Food and this facilitated timely public policy decisions.

In addition to the original Mexican material, we had selected amber grain wheat varieties like *Kalyan Sona* and *Sonalika* from the segregating populations sent by Dr. Borlaug. Farmers in the interior areas of Uttar Pradesh, Bihar and Madhya Pradesh used to refer to the new varieties as "radio varieties" since they had heard about them only through the All India Radio. The media thus helped to convert a small government programme titled, "High Yielding Varieties Programme" into a mass movement. This is why the progress witnessed was revolutionary and not merely evolutionary. Our print media came to the rescue of the country at a time when all the global media and experts had written off India from the point of view of its ability to feed itself. Experts like Paul and William Paddock even applied the triage analysis methodology and came to the conclusion that India cannot be saved from mass starvation and death caused by hunger. It is in this background that the Green Revolution took place and converted India from the position of carrying a begging bowl to becoming a bread basket. The print media and the radio thus served as bright affirming flames in the midst of the sea of despair then prevalent and helped to generate a new confidence in our agricultural capability.

It is now nearly 50 years since the onset of the Green Revolution. There is talk about the need for a second green revolution. However, such a revolution is nowhere in sight. Media faithfully report the Prime Minister's desire for a second green revolution, but have no time or space for discussing why this is not happening. To the financial media, in particular, what matters is Gross Domestic Product (GDP) growth rate, as well as the state of the economy as measured by the situation in the share market. Even this year's widespread drought and the consequent suffering caused to millions of children, women and men were dismissed as unimportant to the economic growth rate, since agriculture contributes less than 18 per cent to the GDP. Since we have enough foreign exchange reserves to resort to large-scale food imports, the media attitude in general seems to be "why bother about farmers and farming?" The fact that nearly two-thirds of our population live in villages and that agriculture constitutes the backbone of their livelihood and survival is shoved under the carpet since it constitutes an "inconvenient truth". No wonder over 40 per cent of farmers interviewed by the National Sample Survey Organisation have expressed a desire to quit farming, if there is another option.

The extensive prevalence of child and adult malnutrition and our anticipated failure to reduce the number of hungry by half by 2015, as stipulated under the first of the UN Millennium Development Goals, receive only a passing mention in the media. The social and economic consequences of pervasive hunger and destitution are hardly highlighted. Even during the Green Revolution days of the 1960s and 1970s, Indira Gandhi stressed in her address at the UN Conference on the Human Environment held in Stockholm in 1972 that the health of the environment will depend on the attention we pay to the basic needs of the poor in relation to food, shelter and work. Only mega calamities like severe flood, drought and tsunami as well as farmers' suicides attract media attention. Journalists like P. Sainath, who has been analysing such issues in depth in the columns of *The Hindu*, are rare.

In 2007, a National Policy for Farmers, based on a draft provided by the National Commission on Farmers (NCF), was placed in Parliament by Sharad Pawar, Union Agriculture and Food Minister. This is the first time either in colonial or independent India that a comprehensive policy for farmers has been developed. All earlier policies were for agriculture and not for the men and women who toil in the sun and rain to feed us. The National Policy for Farmers calls for a paradigm shift from measuring agricultural progress merely in terms of growth rates, to measuring it in terms of the growth in the real income of farm families. The Policy stresses the need for an income orientation to farming both for overcoming poverty in rural India, and for attracting and retaining youth in farming.

The famine of jobs is one of the primary causes of food insecurity in the country. Recent reports that over 5000 persons apply for each clerical job in Railways or Banks is evidence of the growing frustration among educated youth. Agriculture can trigger job-led economic growth, provided it becomes intellectually satisfying and economically rewarding. This will involve the technological upgrading of small farm agriculture and to giving small farmers the power and economy of scale through appropriate group management innovation. This will also call for strengthening the services sector relevant to small-scale farming, such as agri-clinics, agri-business centres, and the Small Farmers' Agri-business Consortium.

Nearly 60 per cent of our cultivated area is rain-fed, with the result our agriculture continues to remain a gamble in the monsoon. There are however new technologies which can help to enhance the yield of dryland crops like pulses and oilseeds by 200 to 300 per cent. Our imports of pulses are increasing, while there is great scope for producing the pulses and oilseeds we need in our rain-fed areas. In the 1960s, media correspondents visited experiment stations and conveyed to readers the excitement of the new genetics. Such visits and reports are rare now.

The Copenhagen Summit has led to a considerable discussion on common and differentiated responsibilities with reference to the containment of Greenhouse Gas emissions. However, there is very little discussion on the potential impact of a rise in mean temperature by 2°C, as agreed at L'Aquilla by G-20 nations. Such a rise in mean temperature will diminish the production of wheat, rice and other crops significantly. Food production will also be affected globally and the price of basic staples will go up. We cannot therefore depend on imports for meeting the food needs of our growing population. Also, global warming will affect rural women more, since they are traditionally involved in the selection of feed and fodder, care of animals and fetching water. The gender dimensions of the impact of climate change are receiving scant attention.

We are producing about 220 million tonnes of cereals to meet the needs of a population of 1.15 billion. While calculating food requirements, we often overlook the needs of farm animals. We have nearly one billion farm animals, including poultry, needing both feed and water. We have to double cereal production by 2050, if we are to meet the needs of the expected human population of nearly 1.8 billion, in addition to meeting the needs of livestock and poultry. Globally also the human population is expected to reach 9 billion by 2050 and global food production will have to go up by at least 70 per cent to meet needs. A couple of years ago when petroleum prices went up, food prices also went up and there were food riots in many countries. The saying "Where hunger rules, peace cannot prevail" is not just a cliché. The media will have to serve as sources of early warning in relation to the emerging food crisis.

Finally, farmers get inadequate and uncertain prices for their commodities. This is why NCF recommended that the minimum support price for rice, wheat and other commodities should be C_2 (i.e., total cost of production) + 50 per cent. Fortunately the Commission on Cost and Prices adopted this formula while recommending a support price Rs. 1080 per quintal for wheat in 2008. I hope the 2009 announcement of Rs. 1100 per quintal of wheat will be suitably adjusted at the time of procurement, taking into account meteorological conditions and the price of inputs. Unless the media assumes a pro-small farmer approach in its reporting, food production will either stagnate or go down. This will obviously affect the country's social stability adversely. It is time that the media resume their active participation in revitalising our agriculture and in safeguarding our food sovereignty.

Chapter 20

Resolving Asia's Contradictions: Growth Versus Inequities

Asia was for a long time known as a "sleeping giant" because of the dichotomy prevailing in most Asian countries between the prosperity of nature and the poverty of the people. Colonial powers came to Asian countries because of the abundant natural and mineral resources of the countries of this vast continent. Asian countries are also rich in their spiritual, cultural and culinary heritage. Beginning with the Industrial Revolution in Europe, the technology divide has been an important factor in the North–South economic divide. Japan has been the earliest country to master new technologies and to bring about a paradigm shift from unskilled to skilled work in its hard-working population. From 1980 onwards, China has mastered new technologies in every sphere of human endeavour, as was evident from the spectacular Olympic Games held in Beijing recently. Many Asian nations have also been making impressive progress in GDP growth in the post-colonial era. Asia is also the home of the Green Revolution in agriculture.

In spite of impressive economic growth and technological capability, the Asian growth story is characterised by serious social inequities. The continent is witnessing many divides such as urban–rural, digital, genetic, gender, social, economic and technological divides. The Asian identity is therefore one of ecstasy and agony. We can be proud of our ancient cultural heritage and more recent technological and athletic achievements. On the other hand, we are confronted with three major groups of inequities.

First and the cruellest form of inequity is the prevalence of widespread maternal and foetal under- and malnutrition, resulting in the birth of babies with Low Birthweight (LBW) (about 2.5 kg and below). Such LBW children are disadvantaged at birth in relation to cognitive abilities and brain development. Thus a child, for no other fault except being born in a poor family, is denied the opportunity for the realisation of his/her innate genetic potential for physical and mental development. This situation is serious in several South Asian countries, where almost one out of every four newborn children is characterised by LBW. Denial of opportunities for intellectual development in the knowledge age in which we live is inexcusable. Hence, there must be a serious effort to ensure that all pregnant women have access to adequate and balanced diet, clean drinking water and environmental hygiene. Elimination of malnutrition-induced inequity at birth should receive the first priority in the struggle for growth with equity.

The second aspect of inclusive economic growth covers all forms of intra-generational equity. These include the goals of literacy, health, nutrition and work for all. Unfortunately, in all these basic minimum needs, inequity prevails to varying extents in different countries of Asia. Many Asian nations, with some notable exceptions like China and Vietnam, are yet to achieve a proportionate reduction in the number of malnourished children, women and men in order for them to achieve the first among the UN Millennium Development goals, viz., reduction in the number of persons going to bed hungry and the number of persons suffering from poverty and destitution to half by 2015. Priorities in development strategies and resource allocation decisions must therefore go to promoting equity in access to nutrition, healthcare, education, sanitation and livelihood opportunities.

A third form of inequity relates to the harm the present generation inflicts on the well-being of the generations yet to be born. Such inter-generational inequity can do greater harm in population-rich countries like China, India and Bangladesh. The most serious among the different categories of inter-generational inequity is anthropogenically-induced climate change. Climate change could result in adverse changes in temperature, precipitation, and sea level, leading to serious droughts, floods and coastal seawater inundation. Sea level rise could lead to the submergence of coastal areas and islands like the Maldives. Other forms of environmental damage like loss of biodiversity and pollution of water have equally serious repercussions.

Thus, inequity at birth, intra-generational inequity in adult life and inter-generational inequity pose a serious threat to sustainable human security and well-being. The poor nations and the poor in all nations will be the worst sufferers since they have limited coping capacity to face the challenges of global warming and environmental degradation. Both unsustainable lifestyles on the part of the rich, and unacceptable poverty on the part of large sections of the population are threats to peace and security. Remarkable advances in many areas of technology like information communication technology (ICT) and eco-technology have opened up uncommon opportunities for ushering in an era where there can be harmony between humankind and nature and also harmony among members of the human family. Let Asia, with a rich repository of traditional knowledge and modern science, show the way.

Chapter 21

From Killing Fields to Smiling Gardens in Northern Sri Lanka

The killing war has ended but the struggle for economic survival is just beginning for the Tamil population of the Northern Province of Sri Lanka. Peace has come but for this to last, the union of hearts is essential. Thus, the Tamil population of Sri Lanka need both peace and bread with human dignity for becoming proud citizens of this beautiful country linked spiritually to India through Buddhism. While such issues have to be settled politically, the immediate need is that of livelihood rehabilitation.

Over 80 per cent of the population of the Northern Province of Sri Lanka, comprising the districts Jaffna, Kilinochi, Mullaithivu, Mannar and Vavuniya, depend on farm enterprises for their work and income security. Prior to the conflict, the Province had a cultivable area of about 300,000 ha, of which over 100,000 ha were under rice. The remaining area was planted with onion, chillies, mungbean, sesame, groundnut, yams, fruits, vegetables and coconut. At the moment, it is not known how much of the area can be farmed, since there has been much damage to the ecological foundations of agriculture — soil, water and biodiversity — due to landmines, protective structures and security measures. Collecting fresh agricultural statistics is a priority task.

The Northern Province was an agricultural paradise before the conflict. Currently, it is estimated that crop production may be of the order of 150,000 tonnes, as against 800,000 tonnes earlier. A large proportion of farms is in the form of home gardens of about 0.1 ha, producing a variety of horticultural and animal products. There is great scope for dairy, goat and poultry farming. There are also large numbers of fisher families along the coast. The Tamil farmers of the Northern Province have been the major custodians of Sri Lanka's food security, much in the same way as the Sikh farmers of the Punjab, who contribute much of the wheat and rice for our public distribution system.

The Northern Province had over 2300 minor tanks and about 50 medium and 11 major tanks prior to the war but most of them now need restoration. Building an irrigation security system is essential for agricultural revival. Forests need to be replanted and rejuvenated. Because of the loss of life and livelihoods, the working population among men has been reduced considerably. The feminisation of agriculture calls for making all agricultural research, education, extension and development programmes gender sensitive. Thus, the pathway to livelihood rehabilitation in this area is agricultural renewal. India, with its vast infrastructure in agricultural research and education, is in the best position to usher in an era of hope in the lives of the Tamil farm and fisher families.

During a visit to Sri Lanka in June 2009, I had the opportunity to discuss issues relating to agricultural research, education and development with the President, Prime Minister and other concerned Ministers of the country. The suggestions contained in this article are based on these discussions.

Strategies for the Wanni Region

The Ministry of Agriculture of Sri Lanka has prepared a strategy for "Vadakkin Vasantham" (Revitalisation of the Northern Province). The President mentioned that because of very high subsidy (96 per cent), chemical fertilisers are displacing compost and other organic manures. An Integrated Nutrient Supply System comprising vermiculture (earthworm), biofertilisers, compost, cereal-legume rotation and green manure crops will help to reduce the dependence on mineral fertilisers. Green manure plants like *Sesbania rostrata* fix nitrogen both in the stem and in the roots. Seeds of such plants should be multiplied and distributed. At the same time, soil health cards giving data on the physical, chemical and microbial status of the soil should be issued to each farm family.

Every calamity presents an opportunity for progress. A new chapter in the agriculture of the Northern Province can be opened up by introducing farmers to conservation farming and integrated pest management and integrated natural resources management. Soil health care and enhancement, and more crop per drop of water techniques developed by the International Water Management Institute will help to increase yield and reduce the cost of production. Post-harvest processing and value addition will help to generate more off-farm employment and increase earning opportunities for women.

India's assistance in strengthening agricultural research training and development will be timely. There is need to move fast, since the internally displaced population urgently need a message of hope. India can also provide the necessary seeds, biofertilisers and biopesticides, farm equipment and post-harvest infrastructure. Good quality seed is a major constraint. Under the Government of India supported programme, Seed Villages can be organised. Immediately, seeds of appropriate varieties and crops could be supplied to help in initiating farm operations. Long-term measures, like taking river water now going to the sea to the Northern Province, are also needed. This will help farmers to take 3 crops a year. The following 5-pronged action plan will help to improve the well-being of the farm and fisher families of the Northern and Eastern Provinces.

Technology: Institutions designed for technology development and dissemination are in a weak state. Besides strengthening existing institutions, it would be useful to develop an Institute for Agricultural Transformation and Capacity Building at Vavuniya and a Fish for All Research and Training Centre at Mannar. These institutes can function on a hub and spokes model, with link centres at different locations in the north and eastern regions. From the beginning, these institutions should concentrate on the development and spread of ecotechnologies by blending traditional ecological prudence with frontier technologies. Farmer-participatory research and knowledge management should be the approach to fostering green technologies like integrated pest management and integrated nutrient supply. Mixed farming involving demand-driven animal enterprises should be promoted to increase farm income and household nutrition security.

Improved post-harvest technology will help to prevent post-harvest spoilage, and add value to primary products. Women-friendly implements and institutional structures like the organisation of Seed Villages will help to accelerate agricultural revival. In view of the damage caused to arable land by landmines as well as the movement of tanks and heavy military vehicles, soil health restoration is an urgent task.

***Techno-infrastructure*:** Seed and grain banks, farm machinery, repair of farm ponds, water harvesting and storage structures, animal sheds, storage godowns, post-harvest handling, rural roads and communication are priority areas for attention and action. Mobile soil testing vans can help to assess the organic matter content and macro- and micro-nutrient status of the soil and to issue soil health cards to farm families. In view of energy shortage, solar and wind energy will have to be tapped in a systematic manner. Similarly, in the case of fisheries, the infrastructure needs of all links in the capture to consumption chain will need concurrent attention.

***Training*:** The pedagogic methodology adopted should be learning by doing. Farmer to farmer as well as scientist to farmer (Lab to Land) learning will help to bridge the gap between scientific know-how and field level do-how. Women farmers can be trained in organising Seed Villages and in the production of the biological software essential for sustainable agriculture, such as biopesticides and biofertilisers. Training in food processing, packaging, storage and marketing will help to promote value-addition and producer-oriented marketing.

***Trade*:** Opportunities for assured and remunerative marketing determine the economic viability of farming. The cost-risk-return structure of farming should guide the fixation of the minimum support price. A market intervention scheme may be necessary during the next few years, until life returns to normal and the damaged rural infrastructure is rebuilt.

***Biovillage*:** Because of the small farm size, multiple livelihood opportunities are essential for achieving a satisfactory level of household income. Hence, increasing on-farm productivity and expanding off-farm employment opportunities should receive integrated attention. The biovillage model of sustainable human security could be the chosen pathway for rural development. This model lays equal stress on the conservation and enhancement of natural resources, improving the productivity and profitability of small farms, and the generation of market-driven non-farm enterprises based on agro-processing and agro-services. Key centralised services could be provided to improve the efficiency of decentralised production.

In the past, India had rendered timely assistance to Vietnam and Kampuchea in getting their agriculture into a high growth trajectory after the end of the war with USA on the one hand, and the termination of the Pol Pot tyranny on the other. Both these countries are now exporters of foodgrains. Similarly, we should extend appropriate help to Sri Lanka, so that the Tamil farm and fisher families are able to regain their original fame and bread basket status.

Chapter 22

Finding Common International Goals

From the beginning of time, science and technology have been key elements in the growth and development of societies. Entire eras have been named for the levels of their technological sophistication: The stone age, the bronze age, the iron age, the age of sail, the age of steam, the jet age, the computer age and the age of genomics and proteomics. We are now on the threshold of the nano age. Unfortunately, the scientific revolution is taking place at a faster pace than our social evolution. As a result, demographic, digital, gender, genetic, technological and economic divides are growing. The rich-poor divide is widening and jobless economic growth — better described as joyless growth — is spreading. Although skin-colour-based apartheid has ended, technological and economic apartheids are appearing.

The world is facing a trilemma — a triple dilemma. Over three billion women and men, struggling to survive with an income of less than US$2 per capita per day, are crying for peace and equitable economic development. Countries in southern Africa, and Ethiopia, Afghanistan and North Korea are in the midst of serious famines. In India, the severe debt burden of small farmers sometimes results in suicides. Two thousand years ago, the Roman philosopher Seneca said, "A hungry person listens neither to reason nor religion, nor is bent by any prayer". Thus, one aspect of the trilemma is the craving for peace and development which is equitable in social and gender terms.

On another side, there is a growing violence in the human heart. Terms like ethnic cleansing and biological and biochemical terrorism are widely used in the media. The revival of smallpox is becoming a possibility. The nuclear peril has again raised its head. Over 30,000 nuclear weapons are stored in the arsenals of major and minor powers. The availability of large quantities of highly enriched uranium increases opportunities for nuclear adventurism.

The third side of the trilemma is the spectacular progress of science and technology, resulting in an increasing technological divide between industrialised and developing countries. Helping to bridge this divide can be an important contribution by advanced educational and research institutions.

In the 1994 Report of the International Commission on Peace and Food, which I chaired, we anticipated a substantial peace dividend following the collapse of the Berlin Wall and the end of the Cold War. No peace dividend has materialised, instead expenditure on military hardware and internal security is increasing day by day, particularly so as a result of the tragic events of 11 September 2001, in the United States and similar events elsewhere.

Contemporary developmental challenges, particularly those relating to poverty, gender injustice and environmental degradation are indeed formidable. However, remarkable advances in information and communication technology, space and nuclear technologies, biotechnology, agricultural and medical sciences, and renewable energy and clean energy technologies provide hope for a better common present and future. Genomics, proteomics, the internet, space and solar technologies and nanotechnology are opening uncommon opportunities for converting the goals of food, health, literacy and work for all into reality. It is however clear that such uncommon opportunities can be realised only if the technology push is matched by an ethical pull. This is essential for working towards a world in which unsustainable lifestyles and unacceptable poverty become features of the past.

Also, there is a growing mismatch between the rate of progress in science, particularly in molecular biology and genetic engineering, and the public understanding of their short- and long-term implications. There is an urgent need for institutional structures that inspire public confidence that risks and benefits are being measured in an objective and transparent manner. Scientists and technologists have a particularly vital role to play in launching an ethical revolution. The Pugwash Movement is an expression of the social and moral duties of scientists to promote the beneficial applications of their work and prevent their misuse, to anticipate and evaluate possible unintended consequences of scientific and technological development, and to promote debate and reflection on the ethical obligations of scientists in taking responsibility for their work. Rabelais said, "Science is but the conscience of the soul". It is the enduring task of our universities, which are the breeding grounds of leaders who will shape our future, to ensure that science and technology are employed for the benefit of humankind, and not its destruction.

We now have a Global Convention on Biological Diversity to help in the conservation and sustainable and equitable use of biodiversity. We need urgently a similar Convention on Human Diversity. While a convention alone will not halt the growing intolerance of diversity — particularly with reference to religion and political belief — it will help foster a mindset that regards diversity as a blessing and not a curse. Both biodiversity and human diversity are essential for a sustainable future. The human genome map shows that over 99.9 per cent of the genomic constitution is the same in all members of the human family. Universities should do more to spread genetic literacy.

It is also necessary to reflect on methods of giving meaning and content to the ethical obligations of scientists in relation to society. The 1999 World Conference on Science in Budapest called for a new social contract between scientists and society. With a rapidly expanding Intellectual Property Rights (IPR) atmosphere in scientific laboratories, the products of scientific inventions may become increasingly exclusive in relation to their availability, with access limited to those who can afford to pay. The rich–poor divide will then increase, particularly with reference to scientific attention and investment. How can we develop a knowledge management system that will ensure that inventions and innovations of importance to human health, food, livelihood and ecological security benefit every child, woman and man, and not just the wealthy? UNESCO could organise a Global Patents Bank for UN Millennium Development Goals. Scientists and technologists from all universities and public research institutions should be encouraged to assign their patents to such a bank, so that the fruits of scientific discoveries are available for the public good. Such a Patents Bank for UN Millennium Development Goals would stimulate scientists to consider themselves as trustees of their intellectual property, sharing their inventions with the poor in whose lives they may make a significant difference for the better.

Over two centuries ago, the French mathematician, the Marquis de Condorcet, a contemporary of Thomas Malthus, said that the human population will stabilise if children are born for happiness and not just existence. The Government of Bhutan has taken the lead in developing a Gross National Happiness Index, based on the economics of human dignity, love of art and culture, and commitment to spiritual values. Making all well-to-do members of the human family regard themselves as trustees of their financial and intellectual property will be essential for fostering a human happiness movement. The 21st century holds great promise for advancing the human condition provided there is an appropriate blend of technology and public action.

I will end with an appeal issued by Bertrand Russell and Albert Einstein contained in the Russell–Einstein Manifesto:

> We appeal, as human beings, to human beings. Remember your humanity and forget the rest. If you can do so, the way is open to a new paradise; if you cannot, there lies before you the risk of universal death.

Can we rid humankind of the nuclear peril and concentrate on harnessing science and technology for achieving the goals of food, water, health and work for all and forever?

Chapter 23

Looking Back and Looking Ahead

Looking Back

I began my research in the field of agro-biodiversity in 1947 at the Indian Agricultural Research Institute (IARI), New Delhi. The experimental material consisted of species and varieties of non-tuber bearing *Solanum* (Family Solanaceae). *Solanum melongena*, the egg-plant (brinjal) belongs to this group. I was amazed at the variety of eggplants, from different parts of India, in the collections made by my supervisor Dr. Harbhajan Singh. The goal of my study was to understand the genetic relationships among non-tuber bearing *Solanums*. In 1949, Professor JBS Haldane visited my experimental field and observed: "I have never seen such variability in quantitative characters as in eggplant; this plant is ideal for studies in the field of quantitative genetics". He further observed that "while Indian farmers are nurturing genetic heterogeneity in their fields as part of their preference for risk-distribution agronomy, scientists seem to be worshipping genetic homogeneity". Genetic homogeneity enhances genetic vulnerability to biotic (pests and diseases) and abiotic (drought, salinity and flood) stresses and this is why in the earlier systems of cultivation, mixed cropping and crop variety mixtures were preferred.

In 1949, I went to the Agricultural University, Wageningen, the Netherlands, to continue my work on Solanaceae, but this time on tuber-bearing *Solanum* species, particularly on potato (*Solanum tuberosum*). The Dutch farmers cultivating potato in the polder lands were facing serious damage to the crop from the golden nematode (*Heterodera rostachiensis*). Professors Dorst and Toxopaeus, with whom I was working, suggested that I should work on breeding potato varieties resistant to the golden nematode. I found from literature that the species *S. polyadenium* from Peru possessed resistance to the golden nematode. This species was in the Commonwealth Potato Collection maintained at Cambridge, UK, by Professor JG Hawkes. I obtained seeds of this and several other species from Professor Hawkes and started crossing them with a popular Dutch potato variety, *Beintije*. Since *S. polyadenium* was a diploid ($2n = 24$), and *S. tuberosum* was a tetraploid ($2n = 48$), I had to double the chromosome number of *S. polyadenium* in order to cross it with the cultivated potato.

The genetic diversity in *Solanum* species fascinated me and I decided in 1950 to go to Cambridge to work on the Commonwealth Potato Collection. From 1950 to 1952, I did extensive research on tuber-bearing *Solanum* species collected from South America and started to unravel the genetic inter-relationships among them. I also traced the origin of the cultivated potato, *S. tuberosum*. This work earned me my Ph.D. degree from the University of Cambridge in 1952.

In November 1952, I was invited by the University of Wisconsin, USA, to join the Department of Genetics in order to assist in the establishment of an Inter-regional Potato Introduction Station at Sturgeon Bay in Lake Michigan, to house the collection made by Dr. Donovan Correll of the US Department of Agriculture. From 1952 to 1954, I undertook extensive gene transfer research from the wild species of tuber-bearing *Solanum*, using several novel cytogenetic techniques. One of the crosses involving the front-resistant species, *S. acaule* from the Lake Titicaca region of Peru–Bolivia border, resulted in the variety Alaska Frostless released for cultivation in Alaska (Swaminathan, 2010).

The work during 1947–1954 on both tuber-bearing and non-tuber bearing *Solanum* species led to my conviction that we should do everything possible to conserve agro-biodiversity for future generations. On my return to India from Wisconsin in 1954, I joined the Central Rice Research Institute, Cuttack, to work on the breeding of high-yielding varieties of rice based on crosses between *japonica* and *indica* strains. The aim was to transfer genes for fertiliser response from *japonica* varieties to *indica*. This programme gave rise to varieties like ADT-27 in Tamil Nadu and Mashuri in Malaysia. There were however several problems like semi-sterility and the breeding of rice varieties with high yield potential had to wait until 1964, when the Taiwan variety, Taichung Native-1 (TN 1), containing the *Dee-gee-woo-gen* dwarfing gene became available. I was also fascinated by the genetic variability maintained by tribal families of the Koraput district in Odisha. In 1954, the Koraput farm families were sustaining nearly 3000 strains of rice but now it has come down to about 300, as a result of gradual genetic erosion. This emphasises the need for *ex situ* preservation, while not relaxing on *in situ*, on-farm conservation.

I joined IARI, New Delhi late in 1954 and initiated work on the breeding of high-yielding varieties of wheat. Dr. B.P. Pal and his associates were then engaged in breeding wheat varieties for resistance to stem, leaf and stripe rusts (*Puccinia sp.*). I tried different methods like crossing the bread wheat (*Triticum aestivun*) with subspecies *compactum* and *sphaerococum* but these crosses yielded dwarf plants with dwarf panicles and consequently had a low yield potential. In 1959, I came to know of the work of Dr. Orville Vogel of the Washington State University, Pullman, USA, in breeding the semi-dwarf winter wheat variety Gaines by incorporating the dwarfing gene from Norin-10 (*daruma*), a variety bred by Dr. Gonziro Inazuka of Japan. Dr. Vogel had given seeds of this material to Dr. Norman Borlaug who was working in Mexico in the breeding of high-yielding and rust-resistant varieties of spring wheat. The history of the introduction of Borlaug's material into India and the subsequent development of outstanding wheat varieties like Sonalika and Kalyan Sona are described in the book *Science and Sustainable Food Security* cited earlier.

The essential point I wish to make is that biodiversity is the feedstock for successful plant breeding. Most of the successful varieties of rice, wheat and other crops may have 50 or more landraces in their pedigree. Because of the availability of genetic variability, a strategy could be developed in the 1960s to checkmate the spread of leaf, stem and stripe rusts in wheat in north India. On becoming the Director of IARI in 1966, one of the first steps I took was to create a Division of Plant Introduction, to strengthen the ongoing work under the leadership of Dr. Harbhajan Singh in the areas of plant exploration, collection and conservation. Both rice and wheat extensive collections were made to preserve for posterity a sample of the genetic variability now existing in these crops. During this period, I initiated a programme for the collection and conservation of rice varieties from the northeastern region of India. This collection, known as the Assam Rice Collection, had over 7000 varieties and proved to be a veritable mine of valuable genes.

On becoming the Director General of the Indian Council of Agricultural Research (ICAR) early in 1972, I initiated steps to set up a National Bureau of Plant Genetic Resources (NBPGR) at the national level and an International Board for Plant Genetic Resources (IBPGR) at the global level through the Consultative Group on International Agricultural Research (CGIAR). I was then Vice-Chair of the Technical Advisory Committee (TAC) to CGIAR. Sir John Crawford of Australia was the Chair of the first TAC set up in 1971. Both NBPGR and IBPGR (now named Bioversity International) have rendered very valuable service in genetic resources collection and conservation. Also, I took steps to establish National Bureaus of Animal and Fish Genetic Resources and later, the National Bureau of Forest Genetic Resources.

During my tenure as Director General of the International Rice Research Institute (IRRI), Los Banos, the Philippines (1982–1988), I initiated enlarging and streamlining the International Rice Germplasm Centre. IRRI now preserves over 100,000 strains of rice. My strategy for conservation was to map the biodiversity hot spots and initiate steps to save the genetic diversity occurring in such endangered habitats. An example is the rice collection made in the interior parts of Myanmar with the help of army personnel since civilians were not allowed to go to some of these areas. The army personnel were trained in genetic resources collection at Yezin. The outstanding rice varieties developed at IRRI under the leadership of Dr. G.S. Khush were the products of an effective use of genetic diversity.

In 1983, I served as a President of the XV International Congress of Genetics held in New Delhi. I chose "Genetic Conservation: Microbes to Man" as the focal theme for the Congress. In my Presidential Address, I suggested that we should establish a global Cryogenic Gene Bank under perma-frost conditions to serve as a "Noah's Ark" in the field of conservation. This proposal fructified when the Government of Norway set up a Global Gene Vault at Svalbard, near the North Pole in 2008. A similar Gene Vault has been set up at Chang La in Ladakh by the Defence Research and Development Organisation of India (DRDO) in 2009. These facilities involve low operational cost and serve as repositories of valuable genetic material. In spite of the growing awareness of the need for conserving biodiversity, its loss is continuing unabated due to habitat destruction, alien invasive species and industrial agriculture. A Biodiversity Literacy Movement is therefore an urgent need.

Cryogenic preservation does not allow evolution. *In situ* conservation involves both preservation and evolution. Therefore, *in situ* conservation and *ex situ* preservation are both important. I assisted the Commonwealth Secretariat and the Government of Guyana in establishing the Iwokrama Rainforest Conservation programme in 1 million hectares of prime rainforest made available by the Government of Guyana. In this programme, as well as in many others with which I have been associated, I introduced the "4C principle", i.e., conservation, cultivation, consumption and commerce. The "4C principle" generates an economic and social stake in conservation. In my report on the Silent Valley Rainforest in Kerala, submitted in 1979, I proposed the development of this unique rainforest and adjoining forests as a Biosphere Reserve.

All over the world there is increasing realisation of the need to have an integrated conservation strategy involving *in situ* and *ex situ* methods as well as community conservation on the lines

I had indicated in my Volvo Prize Lecture (see the following figure).

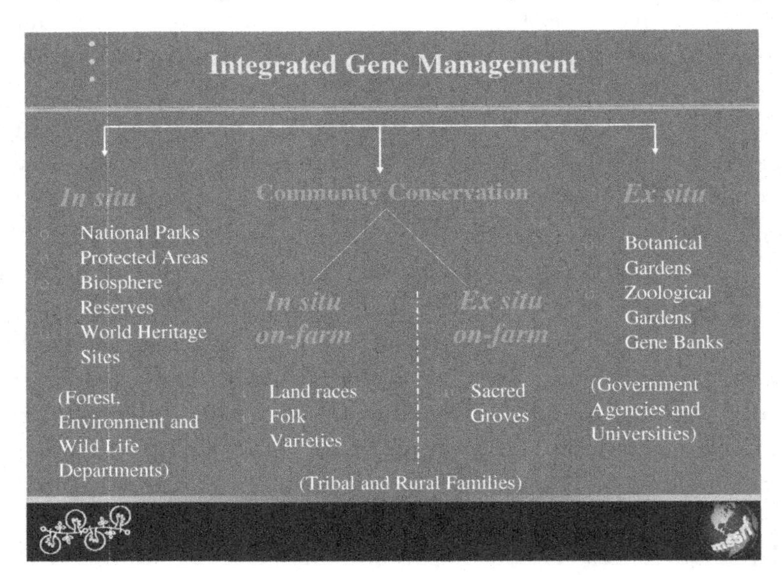

The role of local communities in the conservation and enhancement of biodiversity received inadequate attention and appreciation in the past. Therefore, in the general conference of the Food and Agricultural Organisation (FAO) held in Rome in 1979, I stressed the need for ending the enigma of the poverty of the primary conservers coexisting with the prosperity of those who use their knowledge and material. This led ultimately to the establishment of the FAO Commission on Plant Genetic Resources at a meeting of the FAO Council chaired by me in November 1983. Also, the concept of Farmers' Rights was developed and this was given a legal status under the FAO-sponsored International Treaty on Genetic Resources for Food and Agriculture which came into operation in November 2001.

In the early nineties, the M.S. Swaminathan Research Foundation (MSSRF) started the preparation of draft legislation for the integrated protection of farmers' and breeders' rights. My first draft of such a Bill was supported in an international dialogue held at MSSRF, Chennai in 1994 (Swaminathan, 1995). In 1996, I revised this draft by including farmers' rights in the title of the Act. Thus, was born the Plant Variety Protection and Farmers' Rights Act adopted by the Parliament of India in 2002 (Swaminathan, 1996). Following this, the Plant Variety Protection and Farmers' Rights Authority was set up and the Authority adopted and implemented my suggestion for honouring primary conservers with the Genome Saviour Award.

The community conservation methodology involved promotion of a gene bank (*in situ* on-farm conservation of landraces), a seed bank, a grain bank and a water bank in areas rich in agro-biodiversity. This initiative won for the tribal communities of Koraput in Orissa the Equator Initiative Award at the UN Summit on Sustainable Development held at Johannesburg in 2002. Although the Union for the Protection of Plant Varieties (UPOV) convention has not yet accepted the concept of farmers' rights, it is my hope that my plea that UPOV should become a Union for the Protection of Breeders and Farmers' Rights will become a reality in the near future. Breeders and farmers are allies in the struggle for feeding the ever-growing global population and hence their rights should not only be not antagonistic, but should be mutually reinforcing.

During my presidentship of the World Conservation Union (IUCN) we took steps to prepare a Draft Global Biodiversity Convention. The Draft was discussed and approved at the IUCN Conference which I chaired and which was held at San Jose in Costa Rica in February 1988. Also, the Keystone Dialogues on Plant Genetic Resources held under my chairmanship during 1989–1991, articulated the concept of recognition and reward for primary conservers. The Biodiversity Convention recognises the principles of prior informed consent and benefit sharing. The challenge now lies in getting all nations to accept the concept of farmers' rights and introduce appropriate legislation for the concurrent recognition of breeders' and farmers' rights on the pattern of the Indian legislation.

When I was in the Philippines during 1982–1988, I observed that valuable mangrove forests were being removed for establishing aquaculture ponds. Mangroves serve as bio-shields during coastal storms and tsunamis and promote sustainable fisheries. I therefore helped to establish an International Society for Mangrove Ecosystem (ISME) in 1989 with the help of UNESCO and the Government of Japan.

During my period as Founder–Chairman of ISME (1989–1992), a Charter for Mangroves was prepared. In association with the International Tropical Timber Organization (ITTO), MSSRF organised an international training programme on Mangrove Genetic Resources Conservation. Also, research was started in 1992 on the identification and transfer of genes for seawater tolerance from *Avicennia marina* to rice and other crops by a team of molecular geneticists led by Dr. Ajay Parida. This work has now yielded several salinity-tolerant rice varieties. Recombinant DNA technology helps in transferring genes across sexual barriers and hence no plant or living organism is useless. For example, *Prosopis juliflora*, considered a noxious weed, has provided genes for drought tolerance. The new genetics has brought to an end the era of reproductive isolation of species.

Looking Ahead

My association with biodiversity conservation and utilisation over 63 years has reinforced my conviction that we must do our best to halt genetic erosion, promote biodiversity literacy and make biodiversity conservation everybody's business. Biodiversity is a public good resource and should not be privatised. The Global Convention on Biodiversity and FAO's International Treaty for Genetic Resources both emphasise the need for recognising and rewarding the invaluable contributions of tribal and rural families to biodiversity conservation and enhancement. This is why in delivering the Sir John Crawford Memorial Lecture in Washington DC in 1990, I pleaded for converting UPOV into a Union for the Protection of Breeders' and Farmers' Rights. The farmer is often a breeder and conserver, in addition to being a cultivator. If today, there are nearly 150,000 strains of rice in the world, it is only because of community conservation.

Agro-biodiversity is the result of interaction between cultural and culinary diversity and hence the conservation of cultural diversity and traditional knowledge are equally important. The traditional methods of conservation like sacred groves and temple trees should be revived, since they integrate the spiritual and practical dimensions of biodiversity conservation.

Climate change has reinforced the urgency of conserving traditional crops and wisdom. In October 2010, some 18,000 participants, representing the 193 Parties to the Convention on Biological Diversity, who attended the Nagoya Biodiversity Summit in Japan reiterated the urgency of meeting the unprecedented challenges of the continued loss of biodiversity in an era of climate change. The Strategic Plan of CBD and the "Aichi Target" adopted by the meeting includes 20 major targets organised under five strategic goals that address the underlying cause of biodiversity loss, reduce the pressures on biodiversity, safeguard biodiversity at the ecosystem level, enhance the benefits provided by biodiversity and provide for capacity building.

The Nagoya Protocol included a plan to protect biodiversity by setting targets for 2020. Nations agreed to make 17 per cent of the globe's land area and 10 per cent of coastal and marine areas into protected regions, as opposed to the current levels of 13 per cent and 1 per cent, respectively.

When I was President of IUCN, I used to remark that "conservation without resources becomes just conversation". Fortunately at Nagoya, Japan led the resource mobilisation drive by committing a $2 billion fund for achieving the "Aichi Target" of halving the rate of biodiversity loss by 2020. I hope other countries will follow not only with money but also with emotional, spiritual and political commitment.

When I chaired a committee of the Union Planning Commission in 1980, which recommended the establishment of a separate Ministry of Environment and Forests under strong professional leadership, I had wanted 10 per cent of our land area to be set aside as protected areas. This however was not found feasible by other members of the committee. We should adopt this minimum target and try to achieve it by 2020.

Experience has shown that without education and social mobilisation, regulation alone will not work. I have participated in numerous national and international conferences and workshops during the past 60 years where well-intentioned resolutions and targets have been adopted. Even with reference to the UN Millennium Development

Goal No. 1, i.e., reducing poverty and hunger by half by 2015, progress has been poor in many countries. Unless community understanding and action is combined with national and global resolutions, preventing biodiversity loss will remain a receding goal. For giving local communities space in the management of Biosphere Reserves and National Parks, we should adopt a trusteeship mode, with people and government becoming trustees of these invaluable assets. I got this done in the case of the Gulf of Mannar Biosphere Reserve in Tamil Nadu, by getting its management placed under a Gulf of Mannar Biosphere Trust, with both government and community leaders serving as trustees.

In 2020, there will be a review of the progress made in achieving the "Aichi Target". Considering past accomplishments, there will be disappointment once again unless there is serious effort for making biodiversity conservation a community-led movement. In most of these conferences, administrators, experts and members of civil society organisations participate. They prepare excellent declarations, but these are not followed up by taking the message to those who are the key actors in the conservation movement at the local level. Reaching the unreached and voicing the voiceless will have to become a mandatory public policy in the area of biodiversity conservation.

The tribal women of Koraput are showing how we can convert **biodiversity hot spots** into **biodiversity happy spots** by launching a biohappiness movement involving concurrent attention to conservation, sustainable use and equitable sharing of benefits. I hope their voices of sanity and hope will be heard in the 2020 Conference, since otherwise targets and resolutions adopted at conferences like Nagoya will continue to remain as desirable but unaccomplished objectives.

References

Swaminathan, M. S. (ed.) 1995. *Farmers' Rights and Plant Genetic Resources: A Dialogue*. Madras: Macmillan India Ltd.

Swaminathan, M. S. (ed.) 1996. *Agro-biodiversity and Farmers' Rights*. Delhi: Konark Publishers Pvt Ltd.

Swaminathan, M. S. 2010. *Science and Sustainable Food Security*. Singapore: World Scientific Publishing Co. Pte. Ltd.

Section IV
Towards Furthering Biohappiness

Chapter 24

New Technologies for a Small Farm Productivity Revolution

It is 61 years now in 2014 since the beginning of the new genetics based on the discovery of the double helix structure of the DNA molecule by Watson, Crick and Wilkins. It is also 31 years since the production of transgenic plants started, thanks to the work of Marc Von Montagu, Jeff Shell, Mary del Chilton and several others. The first patent for a living organism went to Dr. Anand Chakraborty who developed through recombinant DNA technology an organism for cleaning up oil spills. The science of molecular genetics has been applied with great benefit in the fields of medicine, industry, environment and agriculture. In the case of medicine, both scientists and consumers have been experiencing many beneficial fall-outs such as new vaccines, insulin and genetic medicine. The major concern in medical genetics is one of ethics, as for example, the application of recombinant DNA technology for reproductive cloning. Therapeutic cloning, on the other hand, has been welcomed. In the case of environmental biotechnology, there is great interest in bioremediation methodologies since there is growing pollution of ground and river water.

In food and agricultural biotechnology, thanks to functional genomics, proteomics, and recombinant DNA technology we are able to address simultaneously the quantitative, qualitative and sustainability aspects of crop production. The rigorous biosafety and environmental safety tests carried out so far have not provided any scientific evidence for adverse impact on human or environmental health. Public good institutions are developing a wide array of valuable genetic combinations, of great importance to increasing small farm productivity and profitability and safeguarding environmental health by reducing the need for applying toxic pesticides. By helping to manage biotic and abiotic stresses, genetically modified crops provide opportunities for avoiding damage by drought, high temperature, flood and sea level rise caused by global warming and climate change. Through an intelligent integration of Mendelian and molecular breeding we are now in a position to avoid threats to sustainable food security. Through a combination of pre-breeding and participatory breeding with farm families, we can combine genetic efficiency and genetic diversity, thereby avoiding the danger of genetic homogeneity.

In the case of agricultural technologies which carry both benefits and possible risks, it is important to have regulatory mechanisms which can help to analyse risks and benefits in an impartial, transparent and professionally competent manner. The same is true in the case of nuclear energy. India is showing the way to the development of an effective regulatory system designed to ensure that the bottom line should be the economic well-being of farm families, food security of the nation, the protection of the environment and the health security of the consumer through the Biotechnology Regulatory Authority Bill introduced in Parliament. Unfortunately the validity of this Bill from the point of view of debate and decision has now expired with the conclusion of the term of the Lok Sabha. This gives the Indian Council of Agricultural Research (ICAR), the Department of Biotechnology (DBT), the Indian Council of Medical Research (ICMR), The Council of Scientific and Industrial Research (CSIR), the University Grants Commission (UGC), Ministry of Environment and Forests (MoEF) and other agencies a wonderful opportunity to go through the text of the Bill once again, taking into account the numerous comments, criticisms and suggestions which have been received, and a get a new Bill prepared for introduction as soon as the new Parliament begins its work. While it may take time to set-up a Parliament-approved National Biotechnology and Biosafety Regulatory Authority, guidelines for safe field testing should be developed. Enforcement of procedures for the release of GMOs for commercial cultivation through the proposed Act may take time but field testing under well-defined safeguards should go on. There are numerous GM varieties in the breeders' assembly line, and they should be tested in the field without further delay. Meanwhile, procedures for their release can be finalised through appropriate legislation.

The Agricultural Biotechnology Committee which I chaired in 2003 and which submitted its report early in 2004 had recommended both a Parliament-approved regulatory agency as well as the necessary infrastructure for conducting all-India coordinated trials with GMOs. Such a special coordinated trial to be organised by ICAR should have as its coordinator an eminent biosafety expert. The necessary precautions, such as the needed isolation as well as demonstration of the importance of refuge, should be undertaken under this coordinated project. Ten years have passed since this recommendation was made and we should lose no further time in implementing it. We should place in position a trial and safety assessment system which answers the concerns of anti-GMO experts and environmental organisations. The present moratorium on field trials with recombinant DNA material is serving as a serious handicap as well as a disincentive in harnessing the benefits of the wide array of transgenic material currently available with various public and private sector research organisations and universities. Many of the GMOs in the breeders' assembly line have excellent qualities for resistance to biotic and abiotic stresses as well as improved nutritional properties. Much of this work has been done in institutions committed to public good. Also much of the work has been done by brilliant young scientists who are getting discouraged because of the lack of a clear official signal on the future of genetic modification.

While urgent steps are needed for putting a widely accepted regulatory system in place, full advantage should be taken of the molecular marker-assisted selection procedures of breeding. The designed goals can be achieved through marker-assisted breeding. Varieties developed through marker-assisted selection are accepted for organic certification. Agriculture is a State subject and it is very important that the State Agricultural Universities and State Departments of Agriculture are involved in the design and implementation of the field trials. It takes nearly 10 years for a new variety to be ready for recommendation to farmers. Therefore, speed is of the essence in organising field trials and gathering reliable data on risks and benefits.

Return from investments in biotechnology research is high. The public sector institutions should accord priority to the development of high-yielding climate-smart and disease-resistant varieties, while obviously the private sector will only produce hybrids whose seeds will have to be bought every year by farmers. Public and private sectors should develop a joint strategy which will help to ensure the inclusiveness of access to improved technologies among all farmers, small or large. The public sector R&D institutions should give high priority to the breeding of varieties which can help farmers to minimise climate and market risks.

There is need for pan-political support for promoting safe and responsible genetic engineering research. Every research institution should have a Project Selection Committee which will examine carefully whether recombinant DNA technology is necessary to achieve the desired breeding goal. In many cases, marker-assisted selection would be adequate for developing a variety with the necessary characters. Recombinant DNA technology should be resorted to only when there is no other way of achieving the desired objective.

Human populations are increasing and may reach 8 billion by 2030. We have no option except to produce more food and other farm products from less land and irrigation water. The population in developing countries is predominantly young, with nearly 70 per cent of the population being below 35 years. Youth will be attracted to choose farming as a profession only if farming becomes both economically rewarding and intellectually satisfying. We will be doing a great injustice to the younger generation of farmers, if we close the gates to the scientific transformation of crop yield and quality and to the conservation and enhancement of the ecological foundations essential for sustainable agriculture like soil, water, forest, biodiversity and climate. Conservation of biodiversity has received added momentum due to biodiversity being the feedstock for the biotechnology industry. To quote Jawaharlal Nehru: 'The future belongs to science and to those who make friendship with science.'

Chapter 25

Overcoming the Curse of Malnutrition

'To a people famishing and idle, the only acceptable form in which God can dare appear is work and promise of food as wages': These were the words of Mahatma Gandhi when he was healing the wounds arising from the Hindu–Muslim divide at Noakhali in 1946. He thus stressed the symbiotic bond among work, income and food security. Fortunately, all Indian political parties are committed to the eradication of hunger and achieving the UN Millennium Development Goals in the area of hunger and poverty elimination. Our former Prime Minister Atal Bihari Vajpayee, for example, said in 2001 on the occasion of the release of the *Rural Food Insecurity Atlas* prepared by MSSRF and the World Food Programme: 'The sacred mission of a hunger-free India needs the cooperative efforts of the Central and State Governments, non-governmental organisations, international agencies and all our citizens. We can indeed banish hunger from our country in a short time.' Prime Minister Manmohan Singh has reiterated this resolve, by stating in addresses such as his Independence Day Address on 15 August, 2008: 'The problem of malnutrition is a curse that we must remove. Our efforts to provide every child with access to education, and to giving equal status to women and to improve health care services for all citizens will continue.' How can we convert this pan-political resolve into practical accomplishment?

The UN Millennium Development Goals adopted by all Member States in the year 2000 represent a Global Common Minimum Programme for sustainable human security and well-being. The first among the 8 goals adopted for accomplishment by the year 2015 relates to reduction in the incidence of hunger and poverty. Unfortunately, recent reviews by the Food and Agriculture Organization (FAO), the International Food Policy Research Institute (IFPRI), the World Bank and other agencies show that far from declining, hunger is increasing, particularly in South Asia and sub-Saharan Africa. FAO estimates that about 75 million more were added to the number of hungry persons during 2007, mainly as a result of rising food prices. It is also becoming evident that hunger is a major cause of poverty. Therefore, anti-poverty programmes have to accord priority to hunger elimination. The economic, ecological and social costs of hunger are high and hence this goal deserves to be on the top of the political agenda and public concern.

In 1981, Indira Gandhi suggested after meeting Vinoba Bhave at the Paunar Ashram in Wardha district that the Wardha district should be converted into a Gandhi district, since Gandhiji spent an important part of his life in this district. She asked me to chair a small group to prepare a blueprint for developing Wardha into Gandhi district. Our first task was to develop a definition for Gandhi district. We defined it as one where no one is below the poverty line and no one goes to bed hungry, not because of doles but because of opportunities for sustainable livelihoods. In other words, bread with human dignity was to be the hallmark of the proposed Gandhi district. At that time, over 80,000 families were below the poverty line and hence specific suggestions were given to raise all these families above the poverty level by creating opportunities for productive and remunerative work. Unfortunately this plan for dedicating Wardha to Gandhiji is yet to be implemented. Even now, it is worthwhile updating this report and transforming Wardha into a hunger and poverty free district dedicated to Gandhiji.

On the occasion of the 60th anniversary of our Independence in 2007, a broad-based Coalition for Sustainable Nutrition Security in India was formed at a meeting held at MSSRF, Chennai. The Coalition comprising national, USAID and UN organisations prepared a report titled *Overcoming the Curse of Malnutrition: A Leadership Agenda for Action.* During a recent discussion of this report, the following 5-point action plan emerged.

Institutional Structures for Public Policy and Coordinated Action in Nutrition

Overcoming malnutrition requires concurrent attention to food (macro- and micro-nutrients, clean drinking water) and non-food factors (like sanitation, environmental hygiene, primary health care, nutrition literacy and work and income security). Achieving the goal of nutrition security for all will need the fusion of political will and action, professional skill and peoples' participation. Such a coalition of policy makers, professionals and citizens will have to start from the village and go up to the national level. The following consultative, policy oversight and monitoring structures are suggested.

- *Panchayat/Nagarpalika*/Local Body

  Council for Freedom from Hunger, established by Gram Sabhas/ Local Bodies, with one man and one woman from each village being trained as hunger fighters.
- *State/Union Territory*

  State Level Committee on Nutrition Security, chaired by the Chief Minister, with all concerned Ministers and representatives of civil society organizations, corporate sector and mass media.
- *National Level*

  Cabinet Committee for Nutrition Security, chaired by the Prime Minister.

A system for horizontal linkages among these three levels of action will have to be developed.

Learning for Success: Converting the Unique into the Universal

Nothing succeeds like success. Therefore, it is important to learn from successful examples of the elimination of malnutrition as, for example, Kerala and Tamil Nadu. Kerala has adopted a universal

Public Distribution System (PDS). A unique combination of ICDS and Tamil Nadu Integrated Nutrition Project (TINP) was launched in Tamil Nadu where TINP identified a community worker to concentrate on families with children between 0 and 3 years of age. From 1982, Tamil Nadu has been operating a universal noon-meal programme for school children, which now covers old age pensioners, destitutes, widows and pregnant women. Support is being extended to nursing mothers. Further, Tamil Nadu has decided to provide rice to the poor at a price of Rupee 1 per kg. This will help to reduce undernutrition substantially. Various indicators of malnutrition show a downward trend in Tamil Nadu. For example, the incidence of severe malnutrition (Grades III and IV) among children aged 0–36 months declined from 12.3 per cent in 1983 to 0.3 per cent in 2000. It would be useful to replicate such effective measures to combat malnutrition in all the States and adopt a universal PDS.

Successful programming experience and health and nutrition evidence show that breaking the curse of malnutrition will require focusing on two important target groups: children under two years of age and women, especially adolescent girls and pregnant and nursing women. The first two years of life represent the critical window of opportunity to break the inter-generational cycle of malnutrition. If this critical window of opportunity is missed, child malnutrition will continue to self-perpetuate and malnourished girl children will continue to grow to become malnourished women who give birth to low birth weight infants who are poorly fed in the first two years of life. Based on successful models, State governments can develop a "Hunger Free State" strategy, with a life-cycle approach to the delivery of nutrition support.

Action at Local Level: Community Food and Water Security Systems

Community food and water security systems including the setting up of Grain, Seed, Fodder and Water Banks can be promoted by local bodies. The food basket should be widened, so as to include a wide range of millets like *ragi*, legumes, vegetables and tubers.

The Panchayat Council for Freedom from Hunger should be assisted with the needed technological and credit support for establishing the Grain, Seed, Fodder and Water Banks. Wherever hidden hunger from the deficiency of iron, folic acid, iodine, zinc and vitamin A in the diet is endemic, food-cum-micro-nutrient supplementation and appropriate and effective fortification approaches (as for example, iodine and iron fortified salt) can be adopted. Every Panchayat/Local Council for Freedom from Hunger could invite a Home Science graduate in the area to serve as a nutrition advisor.

Action at State Level: Coordinating Nutrition Security Initiatives

The State Level Committee on Nutrition Security chaired by the Chief Minister of the State should facilitate the implementation of the numerous ongoing nutrition safety net programmes (national, bilateral and international) in a coordinated and mutually reinforcing manner, in order to generate synergy and thereby maximise the benefits from the available resources. The National Horticulture Mission provides a unique opportunity for applying local level horticultural remedies to major nutritional maladies. Overcoming micro-nutrient malnutrition and intestinal load of infection are urgent tasks. State governments should launch a nutrition literacy movement and set up media coalitions for nutrition security for improved nutrition awareness. Such a media coalition should include representatives of print media, audio and video channels, new media including the internet, and traditional media like folk dance, music, and street plays.

Action at National Level: Mainstream Nutrition in National Missions

At the national level, the most urgent task relates to including nutrition outcome indicators and targets in all major missions in the field of agriculture and rural development. Programmes like the *Rashtriya Krishi Vikas Yojana* (Rs. 25,000 crore), the National Horticulture Mission (Rs. 20,000 crore) and National Food Security

Mission (Rs. 5000 crore) should have a Nutrition Advisory Board, so that cropping and farming systems are anchored on the principle of food-based nutrition security. The National Rural Employment Guarantee Act (NREGA) sites, where mostly illiterate women and men work on unskilled jobs, should have a nutrition clinic operated by a knowledgeable person and a PDS facility. If food is not available at affordable prices at NREGA sites, most of the money earned will go to purchasing staple foods at high cost and under-nutrition will persist.

As a concrete manifestation of the commitment of everyone in the country to achieving Gandhi's goal of food for all, I suggest that the following two steps may be considered, in addition to action at the government level:

- All Members of Parliament and Members of State Legislatures, who are provided with Rs. 1 to 2 crore per year for local area development (MPLAD) should allocate the funds for eliminating malnutrition from their constituencies based on the Gandhi district plan of assisting every family to earn their daily bread. This should be done until malnutrition is totally eradicated from the concerned constituencies.
- All corporate houses should allot the funds available for corporate social responsibility to projects designed to eliminate hunger. Such projects could relate to enhancing the productivity and profitability of small-scale farming and women's self-help groups, as well as to strengthening nutrition safety nets and eliminating leakages in the delivery system.

If the above steps are taken we will be walking our talk and not postpone further erasing the stigma and shame associated with our country being the home for the largest number of malnourished in the world.

Chapter 26

Traditional Knowledge and Modern Science

Knowledge is a continuum. Present-day discoveries often have their roots in prior knowledge. Unfortunately, the Intellectual Property Rights (IPR) regime tends to ignore the contributions of traditional knowledge in the creation of new knowledge. This has led to accusations like biopiracy, plagiarism, knowledge piracy, and so on. The World Intellectual Property Rights Organisation (WIPRO) has hence emphasised the need for recognising the role of traditional knowledge in the growth of contemporary science and technology. Fortunately, the Global Biodiversity Convention adopted at Rio de Janeiro in 1992 and the FAO International Treaty on Plant Genetic Resources for Food and Agriculture (2001) have both stressed the importance of recognising and rewarding traditional knowledge as well as the contributions of rural and tribal families to genetic resources conservation and enhancement through knowledge addition on their practical value. Our national legislations, Plant Variety Protection and Farmers' Rights Act (2001) and Biodiversity Act (2002), have both emphasised the importance of recognising and rewarding traditional knowledge and local agro-biodiversity, which often constitute the backbone of our food and livelihood security systems.

Traditional knowledge has led to the growth of indigenous systems of medicine like ayurveda, unani, siddha, etc. There is a growing awareness of the importance of traditional systems of medicine.

Saving plants for saving lives and livelihoods has become a global goal. Unfortunately however, there is still no methodology for rewarding traditional knowledge, since it involves community recognition, although there are systems in place for providing financial recognition in the field of genetic resources conservation and sustainable use. For example, the Gene Fund provided for in the Plant Variety Protection and Farmers' Rights Act and the Biodiversity Fund provided under the Biodiversity Act can be used for rewarding and strengthening the *in situ* on-farm conservation traditions of local communities. It should be emphasised that while cryogenic *ex situ* conservation leads only to the preservation of specific genotypes, *in situ* on-farm conservation results in both preservation and evolution. New genotypes through mutation and recombination can occur under conditions of *in situ* conservation, while *ex situ* methods involving cryogenic storage can only lead to preservation without loss of viability. Therefore, we should do everything possible to promote *in situ* conservation by recognising and rewarding traditional knowledge and conservation techniques.

Anil Agarwal and Sunita Narain (1997) have chronicled the dying wisdom in relation to water harvesting and conservation techniques developed over the ages. The US National Academy of Sciences has published a series of books such as the *Lost Crops of the Incas*, and the *Lost Crops of Africa*. WHO has been appealing to save plants to save lives, with reference to medicinal plants. Therefore, no further time should be lost in preventing the erosion of traditional knowledge and local biodiversity. Saving plants and traditional wisdom are particularly important to face the challenges arising from global warming and climate change.

There are around 1,500 gene banks in operation today in different parts of the world, providing facilities for *ex situ* conservation for an estimated six million species and varieties. More recently, the Scandinavian countries have established a long-term seed storage facility under permafrost conditions known as the Svalbard Global Seed Vault, which can hold over 6 million seed samples. This will serve like a Noah's Ark in order to preserve for posterity a sample of genetic diversity currently occurring on our planet. However, as already emphasised, cryogenic preservation will not give us the benefit of natural evolution and the further development of new genes and genotypes. This is why recognition of traditional knowledge and traditional conservation ethos is exceedingly important. Sacred groves and sacred trees constituted important methods of conserving economic, ecological and spiritual keystone species. These are also tending to get neglected.

Several steps have been taken in India to recognise and preserve traditional knowledge. A database on indigenous innovations is being kept at the Institute of Management, Ahmedabad, under the leadership of Professor Anil Gupta. The Foundation for the Revitalisation of Indigenous Health Traditions (FRIHT) is also maintaining a database on our heritage of both medicinal plants and traditional medicine and health practices. There are many other initiatives including the Community Gene Bank of MSSRF. We are yet to start a similar programme with animal genetic resources. India is very rich in animal wealth but unfortunately many important breeds including the Vetchur cow of Kerala are now endangered animals. We should institute a Breed Saviour Award to accord recognition to those who are conserving local breeds of cattle, buffalo, sheep, goat, poultry, etc. In the case of poultry, there is indiscriminate killing of native breeds of birds in order to prevent the spread of the H5N1 strain of the avian influenza. In this process, we may lose genes for resistance, in case any of the local breeds possess such genes. Therefore, we should establish an offshore quarantine island in one of the unmanned Lakshadweep group of islands, where in a high security greenhouse testing of local poultry breeds for resistance to H5N1 strain could be conducted. We must strengthen our infrastructure for searching and saving genes, which can help us to overcome emerging challenges caused by both climate change and transboundary pests.

There is need for national and international financing instruments for promoting the conservation of traditional knowledge and endemic bioresources. At the international level, the Global Environment Facility (GEF) is financing measures to implement the Convention on Biological Diversity, the Framework Convention on Climate Change and the Convention to Combat Desertification.

Ten years ago, a Global Crop Diversity Trust was set up in 2004 with an initial capital of US$260 million. The Trust supports information systems for the concerns of agricultural biodiversity, including databases, documentation of collections and the exchange of information through networks.

These international initiatives are important but what is more important is spread of genetic literacy among our population. Every child, woman and man should become aware of the value and significance of traditional wisdom and local biodiversity. This will become easier if there is an economic stake in conservation. We should establish biovalleys in areas rich in bioresources. The aim of the biovalley is to promote an era of biohappiness arising from the conservation and sustainable and equitable use of biodiversity, leading to more jobs and income for the local population. Otherwise both traditional knowledge and native biodiversity may tend to disappear. The power of "the seeing eye and understanding heart" will be evident from the outstanding contribution of the farmer-breeder Joseph Sebastian, whose cardamom variety *Njallani Green* is the ruling species in the Idukki district of Kerala. *Njallani* has helped to improve the productivity and profitability of cardamom and illustrates the power of indigenous knowledge and observation.

Chapter 27

Role of International Years in Meeting the Zero Hunger Challenge

In 2012, the UN Secretary General, Ban Ki-moon, launched the Zero Hunger Challenge Programme, designed to achieve a hunger-free world, with the following words: 'In a world of plenty, no one — not a single person — should be hungry. I invite you all to join me in working for a future without hunger.' At a later meeting of governments, it was decided that 2025 should be the target year for winning the Zero Hunger challenge.

The food security programmes initiated since World War II, both by national governments and international agencies like FAO and WFP, have by and large been designed to address poverty-induced undernutrition. In cereal-based diets, undernutrition also becomes the mother of malnutrition, since calorie deprivation is an important cause for the deficiency of proteins and micro-nutrients in the diet. This is why there is increasing emphasis at both professional and political levels on achieving not merely food security, but both food and nutrition security.

Nutrition security involves physical and economic access to balanced diets, comprising both macro- and micro-nutrients, as well as to clean drinking water, sanitation, primary health care and nutrition education. Deficiencies of iron, zinc, iodine, vitamin A and vitamin B12 in the diet deny nearly two billion people a healthy life. Myers *et al.* (2014) have found that rising levels of CO_2 concentration in the atmosphere resulting in climate change can reduce the content of zinc and iron in crops like rice and wheat, characterised by the C-3 pathway of photosynthesis. The adverse impact of hidden hunger on public health will then increase. Overcoming undernutrition, protein deficiency and hidden hunger hence needs remedial action. Nutrition-sensitive agriculture will help to address all these forms of hunger in an integrated manner. Without mainstreaming nutritional criteria in agricultural cropping and farming systems, the prospect for meeting the Zero Hunger challenge with 2025 as the target year will be slim.

The concept of Zero Hunger was introduced in Brazil in 2003 by the then President, Lula da Silva, who took steps to address in an integrated manner the three major dimensions of food insecurity, namely, availability, access and absorption of food in the body. Availability of adequate food depends on both production and imports, while access is conditioned by purchasing power, and absorption by non-food factors, such as clean drinking water, sanitation, primary health care and nutrition education.

Brazil achieved a substantial reduction in poverty and hunger in a short period of time due to the Zero Hunger Programme. Encouraged by the performance of Brazil, the UN Secretary General has suggested coordinated action in five areas to achieve freedom from hunger by 2025. These are: zero stunted children less than two years, 100 per cent access to adequate food all year round, promotion of sustainable food production systems, 100 per cent increase in smallholder productivity, and zero loss or waste of food.

India has been described in the past as a country where mountains of grains and hungry millions often co-existed, thus emphasising that food production alone is meaningless unless backed up by equity in access. Such a situation may not prevail in the future with the historic passage of India's Food Security Act in 2013, which makes access to the minimum quantity of food required by a person a legal right. The transition from a "ship-to-mouth" existence to right to food with home-grown food marks a proud moment for every Indian. There is, however, no time to relax since there are several serious threats to sustainable food security, such as loss of prime farmland to non-farm uses, water scarcity, climate change and lack of interest among youth to take to farming as a profession. It is in this context that agriculture-centric **International Years** assume importance in generating the necessary political action and public interest in fostering an ever-green revolution in agriculture, which can help increase productivity in perpetuity, without associated environmental or social harm.

2013 was designated by the United Nations as the International Year of Quinoa (*Chenopodium quinoa*). Rich in protein, quinoa has played an important role in food and nutrition security among the Andean people. Thanks to the Quinoa Year, there is now greater interest in under-utilised crops and dying wisdom related to nutrition. The Quinoa Year has also helped to stimulate interest in agricultural remedies to nutritional maladies in different farming systems. Fortunately, the Indian Food Security Act provides for enlargement of the food basket under the public distribution system (PDS) by including a wide range of millets like *ragi,* as well as maize, sorghum, pearl millet (*bajra*), etc. If such a provision is supported by a nutrition literacy programme, agriculture can become the means for both food and nutrition security.

2014 is the International Year of Family Farming. If the year is used for the revitalisation of family farming traditions, with particular emphasis on the empowerment of women and young people, we can help to improve small farm productivity and profitability, on the one hand, and nutrition-sensitive agriculture on the other. Family farming as a way of life and the means to sustainable livelihood helps promote both job-led economic growth and the conservation of biodiversity, thus protecting the ecological and economic foundations of sustainable agriculture. Family farms can play a catalytic role in achieving a shift from food to nutrition security, since nutritional criteria can be integrated in the choice of crops cultivated by family farmers, particularly women. Naturally bio-fortified plants like *moringa*, sweet potato, and breadfruit, and a wide range of fruits and vegetables should find a place in all family farms. Varieties of wheat, rice, and pearl millet, rich in iron, zinc and other vital micronutrients, developed through conventional plant breeding are now available (Swaminathan, 2012). While corporate as well as private farms tend to promote market-driven monoculture of food crops, family farming is generally characterised by diversified cropping patterns, including crop-livestock-fish integrated farming.

The International Year of Family Farming provides a unique opportunity to shift from food to nutrition security at the household as well as national and global levels. The steps needed for achieving such a paradigm shift are: Survey and identification of the major nutritional problems prevailing in an area, introducing appropriate changes in the cropping system to address the deficiencies, and standardisation of measurement tools to estimate the impact of the nutritional interventions on human health. It will be useful to train in each village a woman and a male member of elected local bodies to serve as Community Hunger Fighters. They will have to become knowledgeable of the nutritional deficiencies which can be overcome through a food-based approach. Nutrition-sensitive farming is an idea whose time has come. The International Year of Family Farming can mark the beginning of a "Farming Systems for Nutrition" movement. This will be an effective and economic method of ending concurrently undernutrition, protein hunger and micro-nutrient-deficiency-induced hidden hunger, thereby successfully meeting the Zero Hunger challenge.

2015, dedicated to the conservation and enhancement of soil health, has been named the International Year of Soils. Most of the food we eat today comes from the soil. Land and water use decisions have a feedback relationship. A major threat to food security in the future may be climate change-induced inadequate and uncertain rainfall. There should be integrated attention to soil and water conservation and sustainable use during the Year of Soils, particularly in regions like South Asia and sub-Saharan Africa, where the adverse impact of higher mean temperature is expected to be high and where climate change adaptation and mitigation measures are as yet inadequate. A Soil Health Care movement should be launched during the year.

Finally, 2016 has been selected to highlight the role of pulses (grain legumes) in improving both human nutrition and soil health (through biological nitrogen fixation). Pulses occupy a central place in nutrition-sensitive agriculture, and every effort should be made during the Pulses Year to increase the production and consumption of protein-rich grain legumes, particularly among vegetarians.

It would be appropriate to designate one year during this decade as the International Year of Biofortification and Under-utilised Crops. This will help promote nutrition-sensitive farming practices, and at the same time enable concurrent attention to ending all the three forms of hunger, namely, calorie deprivation or undernutrition, protein hunger, and hidden hunger. This would also help to move from food to nutrition security and thereby address the Zero Hunger challenge in its totality. There is a growing interest among scientists and public health professionals in biofortification as a means of solving the problem of micro-nutrient deficiencies. Many horticultural plants like sweet potato, bread fruit, *moringa* and various berries are rich in micro-nutrients like iron, zinc, vitamin A and vitamin C. Such naturally biofortified plants should find a place in family farming. In addition, food crops can be enriched with specific nutrients like iron, zinc, etc., through plant breeding, as has already been done in crops like rice, wheat, cassava, beans and pearl millet under the CGIAR (Consultative Group on International Agricultural Research) Harvest Plus consortium. Several encouraging examples of the biofortification pathway of enriching staple crops with the desired micro-nutrients were presented at the Second Global Conference on Biofortification held at Kigali, Rwanda in March 2014.

Besides naturally occurring nutrition-rich plants, biofortification of major staples can be achieved through both Mendelian and molecular breeding. A good example of nutritional enrichment through genetic modification is Golden Rice, rich in vitamin A. However, there are still public concerns about the biosafety aspects of crops developed through recombinant DNA technology (GMOs). Fortunately, our knowledge of assessing risks and benefits with reference to GMOs is growing and it is possible to set up regulatory mechanisms which can help to assess biosafety in a scientific and transparent manner (Potrykus 2010).

References

Myers, S. S. *et al.*, 2014. Increasing CO_2 threatens human nutrition. *Nature*. Published online on 7 May, 2014, doi:10.1038/nature13179.

Potrykus, I., 2010. "Regulation must be revolutionised". *Nature* 466, 561.

Swaminathan, M. S., 2012. "Combating hunger". *Science* 338 (1009), 23 November.

Chapter 28

Vision of a Food-Secure India

Agriculture and food security have been intertwined throughout human history. Agricultural growth is critical for improving food security, most immediately by increasing food production and availability. Agriculture helps to grow crops and livestock for food and industrial raw materials and is the main source of calories for the world's population. The availability of food is a necessary but not a sufficient condition to assure food security.

A considerable segment of the world population, particularly women and children, suffer from three major kinds of endemic hunger: Calorie deprivation arising from poverty induced undernutrition; protein hunger caused by inadequate consumption of pulses or milk, fish and meat; and hidden hunger caused by the deficiency of micro-nutrients in the diet.

FAO's most recent estimates indicate that 12.5 per cent of the world's population (868 million people) are undernourished in terms of energy intake, yet these figures represent only a fraction of the global burden of malnutrition. An estimated 26 per cent of the world's children are stunted, 2 billion people suffer from one or more micro-nutrient deficiencies and 1.4 billion people are overweight, of whom 500 million are obese. Most countries are burdened by multiple types of malnutrition, which may coexist within the same country, household or individual. The social cost of malnutrition, measured by the "disability-adjusted life years" lost to child and maternal malnutrition as well as to overweight and obesity, are very high. Beyond the social cost, the cost to the global economy caused by malnutrition, as a result of lost productivity and direct health care costs, could account for as much as 5 per cent of global gross domestic product (GDP), equivalent to US$3.5 trillion per year or US$500 per person. The costs of undernutrition and micro-nutrient deficiencies are estimated at 2 to 3 per cent of global GDP, equivalent to US$1.4–2.1 trillion per year (*The State of Food Insecurity in the World*, FAO 2013).

The internationally accepted definition of food security is that emerging from the World Food Summit of 1996: 'Food security exists when all people, at all times, have physical and economic access to sufficient, safe, and nutritious food that meets their dietary needs and food preferences for an active and healthy life.' This definition reinforces the multidimensionality of food security — availability, access and absorption/utilisation of food.

- ***Availability*** refers to the physical availability of food in desired quantities as determined by production net of feed, seed and wastage plus net imports and draw-down of stocks. As noted in the World Food Summit definition, food security also depends upon the ability to obtain food at all times, including through economic or climatic shocks or non-harvest seasons, as well as the availability of locally acceptable foods, as compared to taboo foods which may be proscribed on the basis of culture, religion, health or economic value.
- ***Access*** is determined by the bundle of entitlements related to people's initial endowments and what they can acquire, especially in terms of physical and economic access to food.
- ***Absorption*** is the ability to biologically utilise the food consumed, which in turn is related to the availability of safe drinking water, sanitation, hygienic environment, primary health care, and nutritional knowledge. This broadening of food security toward nutrition security is a recent evolution.

The problem of hunger is not simply a lack of sufficient quantities of food. The chronic hunger caused by protein and calorie undernutrition is exacerbated by malnutrition (the "hidden" hunger caused by the deficiency of micro-nutrients, which include iron, iodine, zinc, vitamin A, and vitamin B12) and sometimes by human diseases that disable the body's ability to absorb what micro-nutrients it receives. To address such intertwined problems, there must be synergy among national programmes dealing with the availability, access to, and absorption of food. These nutrition security programmes should be based on a life-cycle approach that starts with the "first 1000 days" from pregnancy to 2-years-old, the critical period when stunting can cause irreversible damage, including impaired cognitive ability.

In September 2012, a High Level Panel of Experts to the United Nations (UN) Committee on World Food Security, which I chaired, released a comprehensive report on *Social Protection for Food Security*, with recommendations for combating chronic childhood hunger. One of its recommendations — the concept of a "food security floor" — is particularly worthy of mention. The food security floor recognises that freedom from hunger is a fundamental human right, defining the minimal steps needed for hunger elimination. These include nutrition literacy, clean drinking water, sanitation, and primary health care. In some hunger hot spots of the world where agriculture is the backbone of survival, as in sub-Saharan Africa and South Asia, mainstreaming nutrition in agriculture programmes is the most effective and low-cost method of eliminating malnutrition. This requires greater attention to the net income of smallholder farmers, whose women food producers have particular needs that require specific policies and support.

The challenge before us is the development and adoption of agricultural strategies which can help alleviate poverty and malnutrition. The traditional role of agriculture in producing food and generating income is fundamental, but agriculture and the entire food system — from inputs and production, through processing, storage, transport and retailing, to consumption — can contribute much more to the eradication of malnutrition.

Following the Green Revolution whose thrust was on increasing production through the productivity pathway, agriculture evolved in the decade of the 1970s to include environmental and equity considerations. Sustainable development issues came to the forefront, partly in response to concerns associated with the Green Revolution such as the overuse of agricultural chemicals, the depletion of scarce water resources, and the neglect of farmers and communities in policy-making processes. These concerns encouraged a shift away from a narrow focus on increasing staple food productivity to a more complex perspective on agriculture and rural development. This latter approach coupled intensive agricultural practices with integrated pest management practices, improved water management practices, precision farming, and other tools and techniques that facilitated stewardship of natural resources. Efforts were accelerated to make the Green Revolution not only more sustainable but more pro-poor. New policies, programmes, and investments were specifically designed to integrate rural communities into decision-making processes about their own agricultural and rural development as a way of addressing sustainability along with equity issues. There was growing attention on land reform, especially the equitable distribution of land with secure property rights, access to credit and financial services, and programmes more geared toward small-scale farmers.

In recent years, the medical community has begun to pay more attention to the linkages between agriculture, nutrition and health. There is increasing recognition that agriculture plays a central role in the production, access and use of nutritious and safe food. It also influences other determinants of nutrition, such as access to clean water and sanitation. Health is now considered a major goal of food systems, in part because of the triple burden of malnutrition: Hunger and nutrient deficiencies, as also excess calorie intake that leads to overweight and obesity in many countries.

The Secretary General of the UN, Ban Ki-moon, launched a 'zero hunger challenge' at the Rio+20 conference on sustainable development held in Brazil in June 2012. At a high level consultation held in Madrid, Spain, in April 2013, it was agreed that the world community should commit to a common vision that hunger, food insecurity and malnutrition should be ended by 2025. At these meetings, governments were requested to pay concurrent attention to the following five pillars of the zero hunger challenge: 100 per cent access to adequate food all year round; zero stunted children less than two years of age; all food systems sustainable; 100 per cent increase in smallholder productivity and income; and zero loss or waste of food.

Looking back on India's progress on the agriculture front since 1947, the country has gone through four distinct phases in its agricultural evolution.

- **Phase I (1947–1964):** The emphasis was on the development of infrastructure for scientific agriculture — establishment of fertiliser and pesticide factories, construction of large multi-purpose irrigation-cum-power projects, organisation of community development and national extension programmes, and initiation of agricultural universities. Still, the growth in food production was inadequate to meet the consumption needs of the growing population, and food imports became essential.
- **Phase II (1965–1985):** The emphasis was on maximising the benefits of the infrastructure created during Phase I, particularly in irrigation and technology transfer. The reorganisation and strengthening of agricultural research, education and extension along with the creation of institutions for providing farmers assured marketing opportunities and remunerative prices for their produce led to a quantum jump in the productivity and production of crops such as wheat and rice, a phenomenon christened in 1968 as the Green Revolution.
- **Phase III (1985–2000):** Organisational innovations such as Technology Missions were introduced, the approach involving concurrent attention to conservation, cultivation, consumption and commerce. This period saw a gradual decline in public investment in irrigation and the infrastructure essential for agricultural progress as well as a gradual collapse of the cooperative credit system. Large grain reserves led to a mood of complacency toward agriculture.
- **Phase IV (2001 to present day):** Recent steps seek to revitalise agriculture through several initiatives, including the Mahatma Gandhi National Rural Employment Guarantee Act. Also being

discussed are policies to address the mismatch between production and post-harvest technologies by way of improving of storage facilities.

The Indian enigma is the persistence of widespread undernutrition in spite of substantial progress in agricultural production. Agricultural growth has led to great strides in food production in India, but chronic undernutrition persists.

One part of the solution to this enigma likely involves focusing on crops and livestock that have large nutritional impacts on both farmers and consumers. Another part may involve addressing socio-economic factors that affect the link between agriculture and nutrition, including the distribution of assets, particularly land; the role of women; rural infrastructure; and rural health and sanitation. The Women Farmers' Entitlements Bill of 2011 that I introduced during my tenure as a Member of the Rajya Sabha, was introduced in the Indian Parliament with the aim of establishing women farmers' rights to agricultural inputs, land, water, credit, technology and markets.

I have also been involved in the development of two other policy initiatives to tackle this situation. First, the National Food Security Act (2013) has included nutri-millets in the public distribution system (PDS). These underutilised or orphan crops (referred to officially as "coarse cereals") will be made available at Rs. 1 per kg. This will open up greater market opportunities for these nutritious and climate-smart cereals, thereby providing an incentive to both conserve and cultivate them. The greater the opportunity for remunerative marketing, the greater will be the interest of the farm families in the agro-biodiversity hotspot areas to conserve them. Hence, the widening of the food basket to include millets in the PDS is an important step in converting "hot spots" into "happy spots". Secondly, the Union Finance Ministry provided Rs. 200 crore in the Budget for 2013–2014 for starting a pilot programme on nutri-farms. In such nutri-farms, crops rich in micro-nutrients like iron-rich *bajra*, protein-rich maize, vitamin A-rich sweet potato and zinc-rich wheat will be introduced.

We have to understand that India will remain a predominantly agricultural country for much of the 21st century, particularly with reference to livelihood opportunities. Enhancing small farm productivity and profitability will make a major contribution to reducing hunger and poverty. An integrated crop-livestock-fisheries farming system is the way forward for the country. This calls for an ever-green revolution (i.e., increase in productivity in perpetuity without associated ecological harm), focused on rain-fed farming areas and crops suited to these areas. The technology required has three components:

- **Defending the gains** through soil health enhancement, water harvesting and management, credit and insurance, technology and inputs, and remunerative marketing.
- **Extending the gains** through an appropriate mix of technology, services, and public policies.
- **Making new gains** through improvement in post-harvest technology, agro-processing, genomics and gene pyramiding, and integrated asset reform aimed at equitable land distribution and utilisation of water.

Particular attention is needed to agro-biodiversity hot spots. Predominantly inhabited by tribals, these areas are characterised by culinary and curative (medicinal plants) diversity. Women play a key role here. Over centuries, they have conserved rich genetic variability for public good, at personal cost. More recently, the government, through the National Plant Variety Protection and Farmers' Rights Authority, has started recognising their contributions through the Genome Saviour Award. The following issues are relevant in this context:

1. Commercialisation as a trigger to conservation by standardising methods of creating an economic stake in conservation, thereby helping to improve the economic well-being of the primary conservers.
2. Methods of promoting integrated attention to conservation, cultivation, consumption and commerce, in order to ensure that a

representative sample of existing genetic diversity is preserved for posterity.
3. Strategies for marrying nutrition and agriculture, so that nutri-farms can be promoted.
4. Promotion of farming systems for nutrition (FSN) which can provide agricultural/horticultural remedies to the prevailing nutritional maladies. As an example, the M.S. Swaminathan Research Foundation has designed a FSN initiative, comprising specific steps and is implementing it in the Koraput district of Odisha and Wardha district in Vidarbha. FSN includes carrying out a nutritional survey of the area and identifying the major causes of chronic and hidden hunger, and redesigning the farming system so that specific agricultural remedies are introduced for each nutritional malady, such as the cultivation of bio-fortified crops and crop-livestock integration.

Indian agriculture has now assumed a legal responsibility, since the National Food Security Act 2013 commits itself to a legal access to food to a majority of our population. The Right to Food can be fulfilled only with home-grown food, since international prices are very volatile. Unlike other rights, like the Right to Information which can be redeemed with the help of files, the Right to Food can implemented only with the help of farmers. This is why we have to redouble our efforts in helping farmers to overcome the many challenges they face in producing more food and other agricultural commodities from diminishing per capita land and water resources.

The National Food Security Act 2013 mandates the government to procure wheat, rice and nutri-millets. Such procurement at a remunerative price is the pathway for stimulating interest among farmers to produce more. India is also just beginning to uncover the potential of agri-business, diversification, marketing and exports, as well as increasing the value addition to food production. The country is exploring whether, with proper protections for the poor and vulnerable, commercial agriculture can be a catalyst for economic development. Also, climate change, manifested in adverse alterations in temperature, precipitation and sea level, will add to the problems of farmers and farming. What steps should we take to ensure sustainable advances in agricultural productivity and production?

In my view, we should attend to six key areas to safeguard the stability and sustainability of agricultural production in our country.

First, we should ensure that soil health is not only conserved but improved continuously. This will require concurrent attention to the physics, chemistry and microbiology of soils. Also, we should take steps to ensure that good farmland is conserved for agriculture.

Second, irrigation security will have to be ensured through integrated attention to harnessing rainwater, river and other surface waters, groundwater, treated waste water and seawater. Rainwater harvesting should be made mandatory both in rural and urban areas.

Third, technology and inputs need to be tailored to the agro-ecological and socio-economic conditions under which farmers work. Technology is the prime mover of change and a technology upgrading of agricultural practices via the introduction of biotechnology, IT and proper agricultural mechanisation is essential to attract and retain youth in farming.

Fourth, farmers should receive appropriate credit and insurance support. Credit should be made available at 4 per cent or even lower interest rates as recommended by the National Commission on Farmers (NCF). Insurance procedures should promote group insurance on an agro-ecological basis. Government should promote an **Indian Single Market**, so that agricultural commodities can move across State frontiers without hurdle. This single step would help to eliminate a major cause of price volatility particularly perishable commodities like tomato, onion and potato.

Fifth, assured and remunerative marketing ultimately holds the key for economically viable agriculture. Procurement at the minimum support price (MSP) is the greatest incentive to farm families. The MSP should be C2 plus 50 per cent as recommended by NCF. The WTO regulations may come in the way of providing our small farmers prices that can help to keep them above the poverty line. We should take the stand at WTO negotiations that in the case of countries like India, where over 50 per cent of the population depend for their livelihood on crop and animal husbandry, fisheries and agro-forestry, there should be a **Livelihood Security Box** on the lines of the Green Box provisions, which are being taken advantage of by industrialised countries to provide high subsidies to their farmers. A hunger-free India is a goal which is non-negotiable.

Finally, there is need to give the power and economy of scale to smallholders. This can be in the form of cooperatives, which have been very effective in the dairy sector or producer companies. Group farming through self-help groups can also be promoted. Today, the small farmer has neither the holding capacity nor bargaining power to ensure that he is able to get a reasonable price for his produce. Also, some kind of group cooperation is essential to promote ecologically sustainable production measures like integrated pest management, scientific water management and improved post-harvest management.

The Green Revolution of the 1960s, which helped us to bid goodbye to a ship-to-mouth existence, and launch the Right to Food movement with home-grown food, was the result of a symphony approach with all the main stakeholders participating in a cooperative manner. It is only synergy between technology and public policy that can safeguard the future of our agriculture and help us generate a malnutrition-free India symphony. Technology will help to ensure the ecological sustainability of the production pathway; and pro-nutrition agriculture strategies and public policy will ensure the economic viability of farming through appropriate input and output pricing policies.

2014 is the International Year of Family Farming and we have the largest number of family farmers in the world. As a part of our response to the zero hunger challenge, we should initiate next year "every family farm a nutri-farm" movement. Such a movement should be strategised to

(1) Enhance productivity and profitability of small holdings.
(2) Eliminate protein hunger through the production and consumption of pulses, milk and egg, among others.
(3) End micro-nutrient malnutrition through the use of naturally occurring and biofortified crops.

(4) Mobilise all government programmes to end hunger and issue every family with a Nutrition Entitlements Passbook.
(5) Bring about convergence and synergy among food and non-food factors such as the benefits of the Rajiv Gandhi Drinking Water Mission, Mahatma Gandhi Total Sanitation Programme and National Rural Health Mission, among others.
(6) Integrate the gender dimension in all interventions and pay particular attention to pregnant women and to the first 1000 days in a child's life.

A resource centre should be developed in every village to derive benefit from the agro-biodiversity and nutritional knowledge of tribal and rural women. The major goal of this initiative should clearly be of conserving, cultivating and consuming diversity in order to address the twin challenges of poverty and malnutrition.

2016 will be the International Year of Pulses. During that year we should bridge the demand–supply gap in pulses. Also, we should propose that the United Nations declare one of the years during this decade as the International Year of Underutilised Crops.

To sum up, Indian agriculture has undergone considerable technological and management transformation since 1947, when the country gained independence. The human population, which was about 350 million then has now reached 1.2 billion. There is hence no time to relax. Jawaharlal Nehru said in 1947, "Everything else can wait, but not agriculture", and that message is even more relevant today. It will, therefore, be appropriate that during 2014 we convert Lal Bahadur Shastri's slogan *Jai Kisan* into a reality and focus our energies on realising the goal of a hunger- and malnutrition-free India.

Index

Printed in the United States
By Bookmasters